H. Bode

Pediatric Applications of Transcranial Doppler Sonography

Springer-Verlag Wien New York

Dr. med. habil. Harald Bode
Kinderspital
Roemergasse 8
CH-4005 Basel
Switzerland

With 45 Figures

ISBN-13:978-3-211-82073-5   e-ISBN-13:978-3-7091-8985-6
DOI: 10.1007/978-3-7091-8985-6

Foreword

The measurement of the cerebral circulation in children, particularly in newborns and young infants, has for a long time been high on the list of needs in clinical and scientific pediatrics. The methods available to date have either been too unreliable or unsuitable for use on children. In the course of a research project at the Department of Pediatrics of the University of Freiburg, Dr. Harald Bode has made the first systematic examination of the cerebral circulation of children using transcranial Doppler sonography.

Over 500 children with ages between 0 and 18 years were included in this exhaustive study, documenting Doppler measurements in about 3,000 basal cerebral arteries. Basic reference values were obtained which involved adapting the methodology and available equipment to the special requirements of the pediatrician. Moreover, the influence of biological and physiological factors on these Doppler values has also been considered in addition to those of disease and therapy.

The result is an impressive record of the many applications of transcranial Doppler sonography during childhood. It is not difficult to predict that this methodology will be of lasting value and capable of further development. I hope this book receives the attention it undoubtedly deserves and that the author is able to continue in realizing his fruitful scientific ideas in clinical pediatric practice.

Freiburg im Breisgau, August 1988   Wilhelm Künzer

Acknowledgements

I should like to thank the many people who contributed to the materialization of this book. I am grateful to Professor Wilhelm Künzer, who guided me into pediatric research and always supported and encouraged my work. My gratitude also goes to Dr. H. M. Strassburg, who introduced me to the fascinating field of cerebral Doppler sonography, and to Dr. A. Harders for his valuable advice regarding the application of transcranial Doppler sonography in pediatric neurosurgery. The late Dr. U. Wais gave his patient assistance in the statistical preparation of this monograph.

My special thanks go to Dr. Alec Eden and also the German Research Foundation for kindly providing the Doppler equipment. I also thank Virginia Sonntag-O'Brien for the English translation, Hans-Dieter Hartmann for the pictures, and Sybille Zenzinger for the photography work. The friendly cooperation of Mr. R. Petri-Wieder from Springer-Verlag Wien is gratefully acknowledged. Finally, I am gratefully indebted to my wife for her patience and encouragement all along the way.

Basel, August 1988   Harald Bode

Contents

Contents

Abbreviations

ABP	arterial blood pressure
ACA	anterior cerebral artery
AUC	area under the curve
BA	basilar artery
CBF	cerebral blood flow
CBV	cerebral blood volume
CPP	cerebral perfusion pressure
CVR	cerebrovascular resistance
DSS	Duplex Scan Sonography
HCT	hematocrit
Hz	hertz
ICA	internal carotid artery
ICP	intracranial pressure
KHz	kilohertz
M	mean value
MCA	middle cerebral artery
MEAN	mean peak flow velocity ($= vm$)
MHz	megahertz
mW	milliwatt
n.s.	not significant
p	error probability
PCA	posterior cerebral artery
pCO_2	CO_2 partial pressure
PDA	patent ductus arteriosus
PI	pulsatility index $[= (vs - vd)/vm]$
pO_2	O_2 partial pressure
r	correlation coefficient
REM	rapid eye movement
RI	resistance index $[= (vs - vd)/vs]$
S	standard deviation
SD	ratio vs/vd
SIPH	carotid siphon
SM	ratio vs/vm

TCD	transcranial Doppler sonography
V	variation coefficient
vd	enddiastolic peak flow velocity
vD	averaged enddiastolic flow velocity
vm	mean peak flow velocity
vM	averaged mean flow velocity
vs	systolic peak flow velocity
vS	averaged systolic flow velocity

1. Summary

The cerebral hemodynamics in healthy and ill children are studied systematically for the first time using transcranial Doppler sonography (TCD). The blood flow velocities in the various basal cerebral arteries are determined from the Doppler shifts. Unlike other Doppler techniques, TCD can be used in patients of all ages. It produces reliable results, even without visualization of the region of the brain being examined. The results are comparable with those of other methods.

The TCD examination techniques used for adults are adapted to fit pediatric conditions. Values for Doppler parameters in the basal cerebral arteries are determined and from these reference values are calculated. The author uses these reference values as the basis for evaluating recordings under various physiological and pathological conditions.

The many factors influencing the Doppler parameters include cerebral blood flow, cerebrovascular resistance, and the diameter of the vessel being examined. With reference to the relevant literature, the findings are evaluated as follows:

During the entire childhood period, especially in the first year of life, age has a tremendous influence on the Doppler parameters. Cerebral blood flow probably reaches its maximum around the 6th year of life. A measurable decrease in the cerebrovascular resistance takes place in the first year. The effect of the CO_2 partial pressure on cerebral blood flow and cerebrovascular resistance can be confirmed by Doppler sonography and must be taken into account when recordings are made under pathological conditions.

Repeated Doppler examinations in infancy are comparable provided the state of vigilance of the babies remains constant. The influence of birth weight and gestational age on the Doppler parameters in the first weeks of life shows the dependence of the cerebral blood flow and the cerebrovascular resistance on the weight and stage of development of the brain. The cerebral hemodynamics are affected by a hematocrit that strongly deviates from the norm.

Doppler sonography can provide information on whether the autoregulation of the cerebral hemodynamics is intact or impaired. In newborns, the position of the body, heart rate changes within the normal range, the bilirubin concentration in the serum, and phototherapy have no effect on the cerebral hemodynamics that is recognizable by TCD.

There are various clinical applications of transcranial Doppler sonography. It provides an improved diagnostic tool for a patent ductus arteriosus Botalli requiring treatment. Various disturbances of the cerebral hemodynamics in the presence of perinatal brain damage can be detected.

Although Doppler sonography is as yet unable to quantify an increased intracranial pressure, it can differentiate the pathophysiological processes.

TCD appears to be a rapid and reliable way to diagnose brain death.

Stenoses and occlusions of basal cerebral arteries can be detected non-invasively.

TCD has already become an established method for monitoring certain forms of therapy, such as hyperventilation therapy.

Further applications of transcranial Doppler in children, especially with the aid of continuous recording, are foreseeable in the future.

2. Introduction

2.1. Subject and patients

The functions of the brain are dependent on an adequate supply of substrates and oxygen provided by the cerebral circulation. There are a number of diseases that can lead to disturbances of the cerebral hemodynamics [20, 133, 188, 362, 398].

Numerous techniques to examine cerebral hemodynamics have been developed since Berger [45] and Kety [204] conducted their investigations [48, 120, 133, 181, 209, 215, 231, 320, 331]. The invasiveness and radiation exposure associated with some of these methods limits their use, especially in pediatrics. Other techniques have yielded unreliable results [209, 331].

Only in the last decade has non-invasive Doppler sonography become a standardized technique. From our present knowledge it seems that the method is harmless [337, 374, 375], puts no strain on the patient, and can be repeated at bedside as often as necessary. Results of adult Doppler examinations of the extracranial blood vessels supplying the brain have been available for some time [64, 65, 120, 144, 201, 259, 309, 311, 322].

Since the beginning of the present decade, it has been possible to examine the intracranial blood vessels in infants through the anterior fontanelle [27, 301, 365, 399]. In 1982, Aaslid was able to record Doppler signals from the large basal cerebral arteries using the transcranial Doppler sonographic method (TCD) [1]. This method became quickly accepted for the diagnosis of intracranial vascular disorders in adults [2–5, 21, 160–162, 239, 240, 252, 257, 318, 325–327, 348].

To date, no extensive study of pediatric applications of TCD has been published. The method is particularly suited for children because they have much thinner skulls than adults so that the vessels are more easily insonated. The cerebral hemodynamics can be examined non-invasively and at relatively little expense in all types of disease and regardless of age.

This study is a report on the first TCD examinations of the cerebral hemodynamics in children. The following aspects will be dealt with:
— adapting the examination technique to the various age groups
— validity of the recorded data
— "normal values"

Table 1. Major diagnoses and number of children examined (n)

Diagnosis	n
Perinatal brain damage	17
Patent ductus arteriosus	29
Anemia/polyglobulia	24
Hyperbilirubinemia	20
Cardiac defect and arrhythmia	13
Hydrocephalus	24
Intracranial malformation or space-occupying lesion	12
Brain death	9
Local cerebral circulatory disturbances	11
Bacterial meningitis	12
Vasomotor headache	10
Migraine	12
Miscellaneous	41
"Normal" children	268

— influence of physiological parameters on the results
— findings in various diseases in children
— evaluation of the results according to current concepts of the physiology and pathophysiology of the cerebral hemodynamics in children
— deriving new concepts from the results
— presentation and critical discussion of current and future applications of TCD in pediatrics.

This book is intended as an aid in introducing and applying the new methodology of TCD in pediatrics. The results presented are based on 915 examinations performed in 502 patients at the Department of Pediatrics of the University of Freiburg in West Germany between 1st September 1985 and 31st December 1986. From approximately 2,800 single TCD measurements, roughly 14,000 Doppler parameters could be evaluated.

The age of the children examined ranged from one hour to 18 years; their weight was between 660 g and 98 kg. Table 1 lists the major diagnoses.

2.2. Cerebral hemodynamics

2.2.1. Physical principles

According to Ohm's law, in a rigid tube the flow volume I is proportional to the perfusion pressure P:

$$I = P/R \tag{1}$$
$$(R = \text{resistance})$$

The volume of a viscous fluid flowing through a rigid tube ($=$ flow volume I) at a continuous, laminar flow according to the Hagen-Poiseuille law is:

$$I = (\Pi * r^4 * P)/(8 * \varepsilon * l) \tag{2}$$

(r = radius of the tube, l = length of the tube, P = the difference between the pressure at the beginning and at the end of the tube, ε = viscosity of the fluid).

Between the flow velocity v and the flow volume I the relation

$$v = I/(\Pi * r^2) \tag{3}$$

applies.

From (2) and (3) we obtain

$$v = (P * r^2)/(8 * \varepsilon * l). \tag{4}$$

From (1) and (2) we derive

$$R = (8 * \varepsilon * l)/(\Pi * r^4). \tag{5}$$

Due to the law of the maintenance of mass and to the incompressibility of fluids in the area of a vessel narrowing, the flow velocity increases inversely proportional to the square of the vessel radius:

$$r(1)^2 * v(1) = r(2)^2 * v(2) \tag{6}$$

where $r(1)$, $v(1)$ = vessel radius and flow velocity before and $r(2)$, $v(2)$ = at the vessel narrowing, respectively:

Here static energy (pressure) is converted into kinetic energy (velocity).

When there is laminar flow in a vessel, the fluid flows in coaxial cylindrical layers. The flow velocities increase from the edge of the vessel to the axis of the vessel, resulting in a parabolic velocity profile.

The Reynolds number Re of 2000 represents the boundary between laminar and turbulent flow:

$$Re = \rho * r * v/\varepsilon \tag{7}$$

(ρ = density, r = radius of the vessel, v = flow velocity, ε = absolute viscosity of the fluid).

The foregoing physical laws are limited in their application to cerebral hemodynamics. There are several reasons for this:

1. Blood is an inhomogenous fluid and not consistently viscous.
2. The flow of blood is pulsatile, not continuous.
3. Blood vessels are elastic, not rigid.
4. Blood vessels have numerous junctions and countless collaterals.
5. Cerebral hemodynamics are controlled by autoregulation.

Nevertheless, these physical principles have essential implications for cerebral hemodynamics.

The volume flow (in ml/min) in a vessel is dependent on the radius of the vessel and on the flow velocity within the vessel. This velocity is in turn dependent on the viscosity (hematocrit) of the blood, on the radius of the vessel, and on the perfusion pressure in the vessel. The flow velocity increases at the site of a vessel narrowing. The poststenotic blood flow velocity is less than the prestenotic velocity due to a conversion of kinetic energy into thermal energy (friction) and sound energy (stenotic murmurs).

2.2.2. Physiological factors

Cerebral hemodynamics are chiefly determined by the following factors: the viscosity of the blood, the arterial blood pressure, the intracranial pressure, the diameter and elasticity of the extracranial and intracranial brain-supplying arteries, the size and extent of the collaterals, the resistance of the cerebral vessels, and the autoregulation.

Blood viscosity

The viscosity of the blood as an inhomogenous, non-viscous fluid is mainly determined by the hematocrit [46, 153, 241, 272, 333, 335, 381, 417]. Also involved are the flexibility of the erythrocytes and their ability to aggregate, the viscosity of the plasma, and the flow velocity [153, 333]. The blood viscosity increases considerably in the capillaries with low flow velocity. In vessels smaller than 1 mm diameter the Fahreaus-Lindquist effect counteracts this increase, as the accumulation of erythrocytes in the axis of the vessel causes a severe reduction of the blood viscosity [5, 250].

Perfusion pressure, flow volume, and vessel resistance

The blood from the left ventricle reaches the capillaries via the carotid arteries and the large cerebral arteries, their collaterals, and the arterioles. From here the blood flows through the subdural veins into the pontine veins, which lead into the venous sinuses. The blood then passes through the jugular vein and returns to the right atrium. The large arteries act as blood conductors, the compliance of which enables blood to flow continuously during systole and diastole. Intravasal pressure reduction (resistance) is the major activity of the cerebral resistance vessels (arterioles, capillaries, and pontine veins).

Because of the special construction of the venous vascular system in the brain, the cerebral venous pressure in the pontine veins can be equated with the intracranial pressure. The cerebral perfusion pressure (CPP) is thus the difference between arterial blood pressure (ABP) and intracranial pressure (ICP):

$$CPP = ABP\text{-}ICP \tag{8}$$

The cerebral blood flow (CBF) is defined as:

$$CBF = CPP/CVR \qquad (9)$$

CVR is the cerebrovascular resistance, most of which is effected by the arterioles and the precapillary sphincters.

Autoregulation

In healthy persons older than one year, the CBF remains constant at blood pressure levels between 50 and 150 mm Hg [250]. The mechanism responsible for this is autoregulation. When the CPP drops, the resistance vessels become dilated causing a decrease in the CVR and enabling the CBF to be kept constant. This prevents inadequate brain perfusion. Conversely, an increase in CPP causes the proximal resistance vessels to constrict, accompanied by a subsequent rise in the CVR. The CBF remains constant. This prevents the increased CPP from being transferred to the thin-walled capillaries, which could result in vessel rupture and cerebral hemorrhage.

This biologically useful principle of keeping the CBF constant occurs within just a few seconds [240, 250]. The mechanisms responsible for this are said to be myogenic and metabolic factors [250].

Physiological influences

Numerous studies are available on the physiological influences on global and regional CBF [250]. CBF has been seen to increase during mental activity and emotional excitement [250] and to decrease when the subject is eating [90, 314]. The CBF is higher during rapid eye movement (REM) sleep than in the nonrapid eye movement (NREM) period [84, 195]. These findings illustrate the dependence of the cerebral hemodynamics on the supply of substrates, that is, on the cerebral activity.

Metabolic-chemical influence

Carbon dioxide: The CO_2 partial pressure (pCO_2) determines the degree of CBF [17, 18, 80, 89, 158, 159, 163, 224, 226, 257, 319]. The CVR decreases when the pCO_2 rises and increases when the pCO_2 drops. An S-shaped CO_2 response curve has been determined [163, 319]. Changes in the pCO_2 have no substantial effect on the CBF below a level of approximately 20 mm Hg (minimum flow) and above a level of 70–80 mm Hg (maximum flow).

In the mean range the CO_2 response curve takes a hyperbolic course [250, 257]. The CBF rises in the physiological range by about 4% per mm Hg increase in the arterial pCO_2 [226]. The steepness of the CO_2 response curve depends on the blood pressure [163]. In the presence of chronic hypercapnia there is either only very little or no CO_2-reactivity.

Oxygen: Acute hypoxia (arterial pO_2 below 60 mm Hg or venous pO_2 below 28 mm Hg) causes the CBF to increase. The autoregulation is disturbed in severe hypoxia [118].

Other factors: Arterial hydrogen ion concentrations have no effect on the cerebral hemodynamics [250]. The extracellular pH level in the vicinity of the cerebral arterioles is important in the regulation of the vessel size [250].

The perivascular vegetative nerves probably play a role in the distribution of blood in the brain. They also regulate the steepness of the CO_2 response curve [250].

2.2.3. Normal values

The healthy adult has a CBF of about 50–60 ml/100 g brain tissue/min [204, 226, 250]. Approximately 700–900 ml of blood flows through the brain per minute, representing 15–20% of the cardiac output. The flow in the different areas of the brain varies considerably [225]. The cerebral blood volume (CBV) in the adult is about 130 ml, with a mean circulation time of 8 seconds [250].

In a 6 year old the CBF is about twice as high as it is in an adult [47, 202]. In the literature the normal values for a newborn are stated as 30–60 ml/100 g brain tissue/min [209]. This corresponds to roughly 30% of the cardiac output [78]. In the premature infant the values are lower (20–48 ml/100 g brain tissue/min) [147, 149, 230, 268]. In the term newborn the CBF increases during the first days of life from 36 to 41 ml/100 g brain tissue/min [275]. According to Wenner's calculations the CBF of the entire brain rises between the 1st and 12th months of life to 1.7 times that of a neonate; the CBF of the cerebral cortex to 3.6 times [409]. In adults the circulating CBF per minute corresponds to about half of the cranial volume. This value is lower in infants and in 6 year olds it is higher [78, 147, 149, 181, 250, 410].

The time of cerebral circulation is said to be shorter in children, with blood flow velocities higher than in adults [344].

2.3. Techniques of examining cerebral hemodynamics

The principles, possibilities, and limitations of the various methods for examining cerebral hemodynamics in children are only be briefly presented in this book. More detailed information can be found in elsewhere [133, 209, 331].

Angiography

Cerebral angiography essentially provides a morphological visualization of the brain-supplying and intracranial vessels. With the aid of special techniques the cerebral circulation time and the cerebral blood flow can be estimated [171, 181]. Because of its invasiveness and its possible side effects [259], strict criteria are to be adhered to in selecting patients for angiography. Nevertheless, the importance of the method, also in children, is indisputable [91, 171, 215, 287, 344, 393, 394].

Xenon-133-clearance technique

This technique is a modification of the nitrogen dilution method [31, 202, 204]. The radioactive carrier substance Xenon-133 is administered intra-arterially [244, 245], intravenously [147, 149] or by inhalation [268]. The mean cerebral blood flow is determined quantitatively (in ml blood per 100 g brain tissue per minute). The differences between gray and white matter in the circulation can be assessed. The regional blood flow can be measured in the brain surface areas [225].

There are various disadvantages associated with the Xenon-133-clearance technique. To obtain exact results the substance must be given intravenously or, preferably, intraarterially. The invasiveness and the use of the radioactive carrier limit its applicability and make follow up examinations impossible. The method also has technical drawbacks. The complicated and time-consuming measurements are influenced by distribution phenomena in the tissue and are disturbed by the extracerebral circulation. The different regions of the brain contribute to the measurements to various degrees. Another disadvantage is that for ethical reasons values from normal subjects cannot be obtained [133, 149, 209, 225, 331].

Vein occlusion plethysmography

Jugular vein occlusion plethysmography [78, 86] measures the change in the size of the head after both jugular veins have been briefly compressed and supplies indirect information on the CBF. However, this method also had disadvantages. The results are dependent on the elasticity of the skull, so that it can only be employed in infancy. Occlusion of the jugular vein causes an increase in intracranial pressure. The compression of the carotid arteries, which often occurs at the same time, results in a reduction of the blood flow to the brain. In many cases the CBF is underestimated [209, 331].

Electrical impedance methods

The human head forms a resistance to the passage of electrical current. This resistance is dependent, among other things, on the volume of blood, so that it changes analogously to the pulse [134, 186, 231, 266, 283]. This principle is applied in studies of cerebral hemodynamics in children [407]. Impedance methods are less costly and time-consuming than clearance techniques and allow for continuous long term recording. However, the measurements cannot be easily reproduced and are falsified by the extra-cerebral circulation that makes up as much as 20% of the entire impedance. The results do not correlate well with the CBF [209, 331].

To date, sufficient data are lacking with regard to measurement accuracy and validity for any of the last three methods.

Positron-emission tomography

Pre-treated radioisotopes are injected intravenously. After they become concentrated in the tissue they begin to emit positrons. The reaction of the positrons with electrons in the tissue causes gamma rays to form [411], which are recorded by detectors from outside the skull [168, 401]. CBF, oxygen consumption of the brain, brain and receptor metabolism can be examined. Unfortunately, the necessity of using radioactive substances makes this interesting method very complicated and expensive. It has hardly ever been applied in children [331].

Magnetic resonance imaging (MRI)

With MRI the magnetic properties of certain cell nuclei are examined in a strong magnetic field. Apart from the morphological visualization of brain structures [16, 143, 229], information concerning the circulatory conditions can also be obtained [94]. This method, however, also requires expensive equipment and long examination times, putting restrictions on its use [331].

Autoradiography

Autoradiography is a technique used in animal experiments by which radioactive labeled macromolecules are applied intravenously to close some of the smallest capillaries. The global and regional CBF can be calculated from the location and the intensity of the emitted radiation [154, 320].

Doppler sonography

Our present state of knowledge indicates that ultrasonic Doppler examinations are harmless, as long as the prescribed boundaries for ultrasonic power and insonation time are observed [337, 374, 375]. Because it is non-invasive, it can be repeated as often as necessary. The velocities of the blood flow in the vessels are measured. Although quantitative measurements of the CBF cannot be made, relative changes can be documented. In addition, the resistance in the vascular system can be estimated. Unlike the above techniques, Doppler sonography requires that the examiner is present during the examination. The results are strongly operator-dependent [34, 39]. In the following chapter Doppler sonography will be discussed in detail.

2.4. Introduction to Doppler sonography

2.4.1. History

Inspired by observations made by the English astronomer James Bradley (1693–1762) [57], Christian Doppler (1803–1853), professor of mathematics and geometry, in his article entitled "Über das farbige Licht der Doppelsterne und einiger anderer Gestirne des Himmels" ("On the colored light of the double stars and certain other stars of the heaven") published in 1842, described the principle which was later to bear his name [103]. A few years later, Buys Ballot (1818–1890), a Dutch mathematician, was able to confirm Doppler's theory in experiments he conducted using the Dutch railroad with horn players as sources of sound waves [70]. The Doppler effect was the change observed in the wavelength of sound (and other) waves due to the relative motion between the wave source and the observer. In 1918, Langevin (1872–1946) was able to obtain electrical oscillations in a quartz crystal and thereby generate ultrasonic waves [72, 411].

Satomura in 1959 [340] and Franklin in 1961 [117] applied these principles to measure the blood flow velocities in peripheral and extracranial brain-supplying vessels. Since then, Doppler sonography has been chiefly used in angiology and neurology [64, 65, 120, 201, 259, 309, 311, 322]. It is also applied for blood pressure measurements [105, 124, 208, 259, 342], in cardiology [112, 141, 382], and in obstetrics [150, 372, 389]. Recently, there have been approaches to measuring the flow volume in the large cervical vessels [145, 279, 384].

For a long time Doppler sonography was not used to examine the intracranial vessels because it was considered that the ultrasonic waves were attenuated by the skull. Although back in 1965 Freund [119] was able to detect the pulsation of some intracranial vessels in the one-dimensional ultrasonic scan (A-scan), technical problems interfered with the recording of a usuable Doppler signal.

The ultrasonic examination through the skull in children also remained confined to one-dimensional methods for some time [278]. At the end of the 1970's it became possible to achieve a two-dimensional visualization (B-scan) of the intracranial structures in infants through the acoustic window of the anterior fontanelle [25, 44, 51, 53, 54, 79, 98, 233, 292]. The pulsations of the large cerebral arteries could be qualitatively evaluated.

In 1979, Bada introduced the Doppler sonographic examination of intracranial arteries into neonatology with the continuous wave (CW) technique [27]. In the early 1980's, the combined application of the B-scan technique and pulsed Doppler sonography (=duplex scan technique) for the examination of intracranial vessels in newborns was described [190, 191, 365, 280].

A major breakthrough occurred in 1982, when for the first time Aaslid measured flow velocities in the vessels of the circle of Willis through the skull using transcranial Doppler sonography (TCD) [1]. In just a few years this method was introduced into the diagnosis and therapeutic control of intracranial vascular disease in adults [2, 4–6, 21, 160–162, 239, 240, 257, 318, 327, 348, 412].

2.4.2. Principle

High frequency ultrasonic waves penetrate and scatter in tissue. When they strike interfaces having different densities, some are reflected back to the source. When ultrasonic waves are backscattered from moving red blood cells, the waves are shifted in frequency [259, 411]. The transmitter and receiver of the ultrasound are located in the same Doppler probe. When the blood cells move toward the probe, the frequency of the received sound is higher, whereas the frequency is lower than that of the emitted ultrasound when the blood cells move away from the probe.

With the ultrasound frequencies commonly used in medicine (2–10 MHz) the frequency shift, also known as the Doppler shift, lies within the audible range. The examiner hears it as an acoustic signal. The Doppler shift can be further processed and recorded by means of special techniques. Along with the acoustic evaluation, the frequency spectral analysis plays a particularly important role today. (For details on the technical problems associated with Doppler signal processing see [169, 187, 198, 290].)

The Doppler shift is expressed by the following formula:

$$F = 2 * Fo * v * \cos\alpha/c \tag{10}$$

where F = the Doppler shift (Hz), Fo = the mean frequency of the emitted ultrasound, v = the blood flow velocity (cm/sec), α = the angle between the direction of the transmitted sound beam and the axis of the blood flow, and c = the velocity of sound in tissue (= 1,550 m/sec).

Thus, the Doppler shift is proportional to the blood flow velocity when the emitting frequency and angle α remain constant. In the resolution of equation 10 vor v, the effect of the emitting frequency of the instrument on the recorded result is eliminated. The blood flow velocity v (in cm/sec) corresponds better to the physiological meaning of the recorded value than the Doppler shift F (in Hz) [207].

Equation 10 also demonstrates the influence of the angle α on the Doppler shift F. A maximum value is found at α = 0 degrees; at α = 90 degrees the Doppler shift is zero. When α = 0, 10, 20, 30 degrees, $\cos \alpha$ is 1, 0.98, 0.94, 0.87, respectively. The measuring error caused by the angle of insonation is then 0, 2, 6, 13%.

2.4.3. Techniques

a) Continuous wave technique

In the continuous wave technique [27, 146, 260, 301] an ultrasound beam is continuously transmitted from an oscillating piezo element. The back-scattered ultrasound is continuously received by another piezo element. Signals are measured from the entire penetration depth of the ultrasonic waves. Unlike bidirectional devices, unidirectional instruments are not able to differentiate whether the blood flow is moving towards or away from the Doppler probe.

b) Pulsed (= range gated) technique

With this technique ultrasonic pulses of only millisecond duration are transmitted intermittently. By varying the receiving time, the receiving crystal, which alternatively acts as the transmitter, can register the back-scattered signals from a defined area at a determined depth (= sample volume). The instruments that function according to this technique can be operated undirectionally or bidirectionally.

Duplex scan technique

This method combines the two-dimensional ultrasonic examination technique (B-scan) with pulsed Doppler sonography. Under visual control in the B-scan, the sample volume is placed in the region to be examined by Doppler sonography.

Transcranial Doppler sonography

When ultrasonic waves strike the skull, most of them are reflected or absorbed. The higher the ultrasonic frequency and the thicker the bone, the less acoustic energy can penetrate. The temporal bone is relatively thin and allows ultrasonic frequencies of 1–2 MHz to penetrate through individual "windows".

The bone itself acts as an acoustic lens which focusses the ultrasonic beam or scatters it. Further, the beam may be refracted, depending on the thickness of the bone and the angle of insonation. Such phenomena effect substantial changes in the emitted and reflected ultrasonic waves. The geometric accuracy of the ultrasonic instrument (e.g. the position and shape of the sample volume, depth of focus) and its resolution are also affected [151].

Despite these difficulties, transcranial Doppler sonography is able to obtain usuable Doppler signals and reproducible measurements from definable intracranial vessels in about 95% of adult patients [21, 161].

2.4.4. Doppler parameters

Doppler sonography measures the flow velocities in the arteries that change according to the cardiac phases and the relatively constant flow velocities in the veins. The Doppler signal contains a spectrum of frequencies corresponding to the parabolic velocity profile of the blood cells in the vessels. The flow velocities can be calculated from the Doppler shift according to equation 10:

$$v = F * c / (2 * Fo * \cos\alpha) \tag{11}$$

This monograph will be confined to Doppler investigations of the arterial system. The following parameters are defined (Fig. 1):

— From the envelope curve of the Doppler frequency spectra which corresponds to the peak flow velocity measured in the course of time: systolic peak flow velocity (vs), mean peak flow velocity (vm), enddiastolic peak flow velocity (vd).

— From all of the registered Doppler frequencies, averaged systolic flow velocity (vS), averaged mean flow velocity (vM), averaged enddiastolic flow velocity (vD).

— The following are computed:

The resistance index according to Pourcelot [311] (RI):

$$RI = (vs - vd)/vs \tag{12}$$

The pulsatility index according to Gosling [144] (PI):

$$PI = (vs - vd)/vm \tag{13}$$

The systolic/diastolic ratio (SD) [109]:

$$SD = vs/vd \tag{14}$$

Further,

$$RI = 1 - 1/SD \tag{15}$$

and

$$SD = 1/(1 - RI) \tag{16}$$

The systolic/mean ratio (SM):

$$SM = vs/vm \tag{17}$$

The area under the Doppler curve (AUC):

$$AUC = vm * (t_2 - t_1) \tag{18}$$

On the envelope curve of the Doppler spectra, t_1 is the beginning and t_2 the end of a pulse wave.

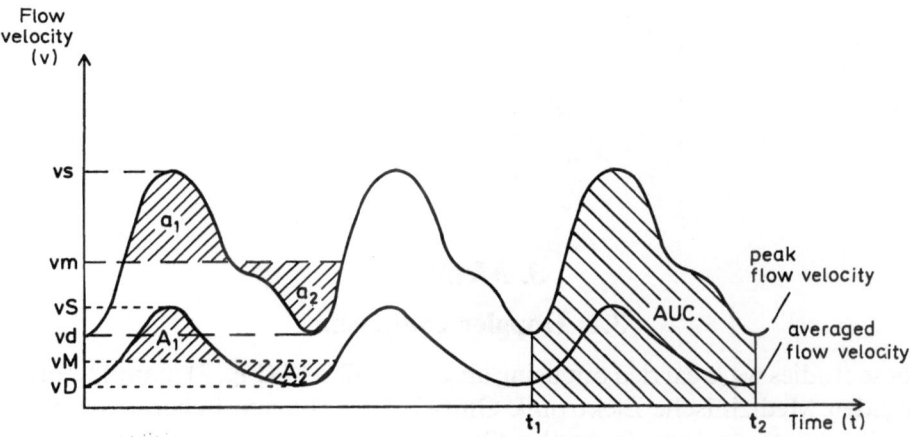

Fig. 1. Parameters of the Doppler signals: $a_1 = a_2$, $A_1 = A_2$
(other abbreviations are explained in the text)

3. Methods

3.1. Doppler equipment

These studies were carried out using an EME TC 2-64 (Fig. 2) manufactured by Eden Medizinische Elektronik GmbH, Ueberlingen, Federal Republic of Germany (on loan from the German Research Association—DFG) which has the following technical features [109]:

— emitting frequency 2 MHz, pulsed
— pulse repetition frequency 5–10 KHz
— pulse duration 13 microseconds
— axial gate width 13 microseconds
— acoustic intensity 100 mW/cm^2 (max. spatial peak-temporal average) adjustable in steps 10–100%
— sound transmitting area of Doppler probe 1.5 cm^2, focussing of the sound beam with a polystyrene lens at a distance of 5 cm as measured in a water bath, where the axial resolution is 5 mm, lateral resolution 8 mm
— size of the sample volume: approx. 4×9 mm at a depth of 5 cm

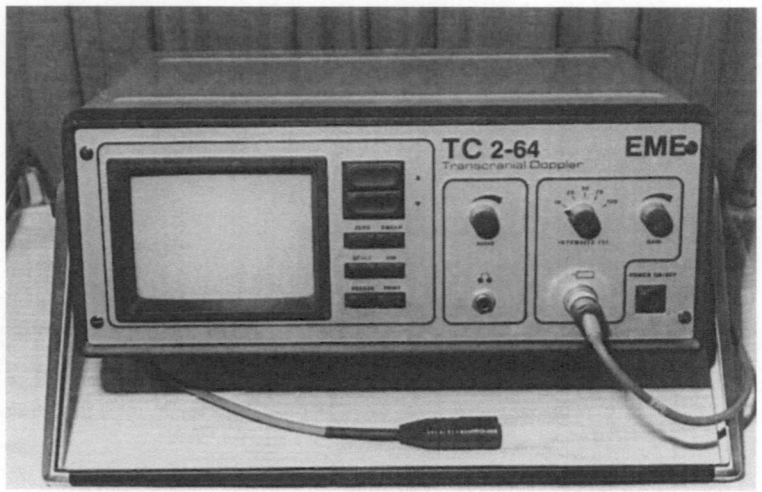

Fig. 2. EME TC 2–64 with Doppler probe

— measuring depth 25–155 mm, variable in 5 mm steps
— low pass filter 10 KHz, high pass filter 150 Hz, so that velocities under 6 cm/sec are not measurable
— maximum measurable velocity in bidirectional recording ± 200 cm/sec, in unidirectional measurement 400 cm/sec
— signal processing by means of frequency spectral analysis (Fast Fourier Transformation) with a frequency resolution of 64 points and a temporal resolution of 13 milliseconds.
— Scales: bidirectional 0–50, 0–100, 0–150, 0–200 cm/sec; unidirectional 0–100, 0–200, 0–300, 0–400 cm/sec.

The frequency (f) can also be measured in KHz, which is done according to the following equation:

$$f = v/0.039. \qquad\qquad (19)$$

From the envelope curve of the Doppler spectra (Fig. 1), vm (= MEAN) and the quotient vs/vd = SD are automatically determined. The examiner can determine vs and vd with a cursor. The optimal Doppler signals, that is, the clearest ones and those free of interference, are picked up by audiosignal analysis over loudspeakers or headphones and are displayed on the video monitor. Measurements lasting 3, 5, 8, 12, 15 seconds can be frozen and documented with an Epson FX 80 matrix printer.

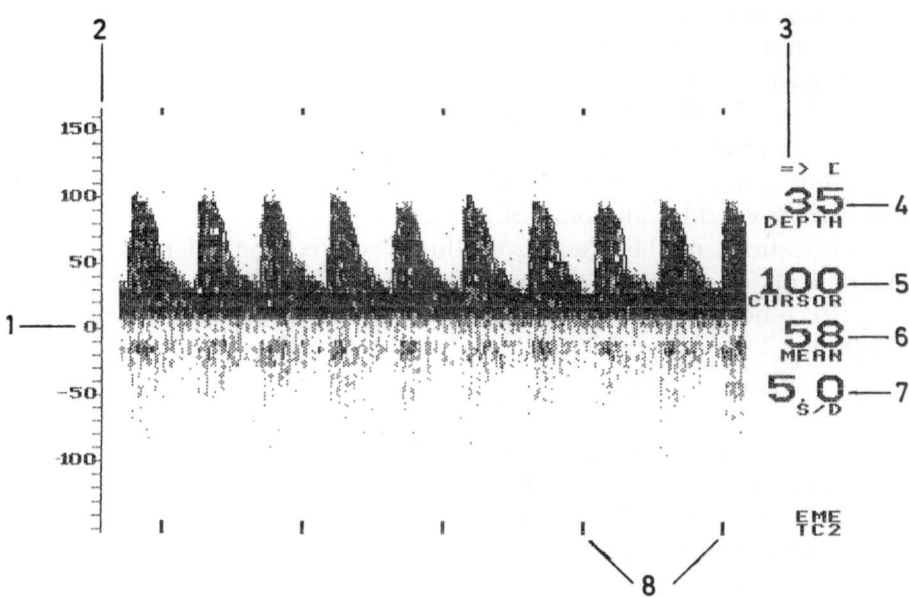

Fig. 3. Monitor print-out from EME TC 2–64 (see text for details)

The following data can be observed (Fig. 3):

1. Zero line: bidirectional measurement in the middle of the monitor; unidirectional measurement on the bottom of the monitor;

2. Scale: flow velocity (cm/sec) or Doppler shift (Hz);

3. Flow direction: if the arrow points to the right, flow toward the Doppler probe is shown above the zero line and flow away from the probe below the zero line. The arrow pointing to the left indicates the opposite;

4. Depth: distance in mm between probe surface and sample volume;

5. Cursor: allows manual measurement of Doppler signal when monitor image is frozen;

6. MEAN (= vm): the mean peak flow velocity calculated automatically by the instrument from the envelope curve of the Doppler spectra during each sweep;

7. SD: the ratio vs/vd of each sweep determined automatically by the instrument;

8. Duration of measurement: one second between two markings.

3.2. Further improvements

Since carrying out this study, the author has used a more recent model of the Doppler equipment—the TC 2–64 B Multifrequency Transcranial Doppler (Eden Medizinische Elektronik GmbH, Ueberlingen, Federal Republic of Germany) which also allows examinations to be performed at transmitting frequencies of 4 and 8 MHz and the Doppler data to be displayed in the same way as for the 2 MHz referred to above. The 4 MHz probe is particularly valuable for the examination of neonates, where penetration of the skull is not a major problem. The use of this higher frequency enables the sample volume to be moved in smaller increments (2.5 mm) and permits a minimum measuring depth of 7 mm, which is more suited to the cerebrovascular anatomy of neonates.

Modifications in the filtering of this newer equipment now permits measurements of velocities of down to 1 cm/sec. Gosling's index of pulsatility (PI) is now displayed at the end of each sweep instead to the quotient vs/vd (SD).

3.3. Recording technique

The anatomical position and variability of the large basal cerebral arteries are given in the literature [91, 215, 219, 287, 344, 393]. Several positions are used to examine the basal cerebral arteries:

1. transtemporal
2. transorbital
3. suboccipital through the foramen magnum
4. high-frontally and through the anterior fontanelle (in infants).

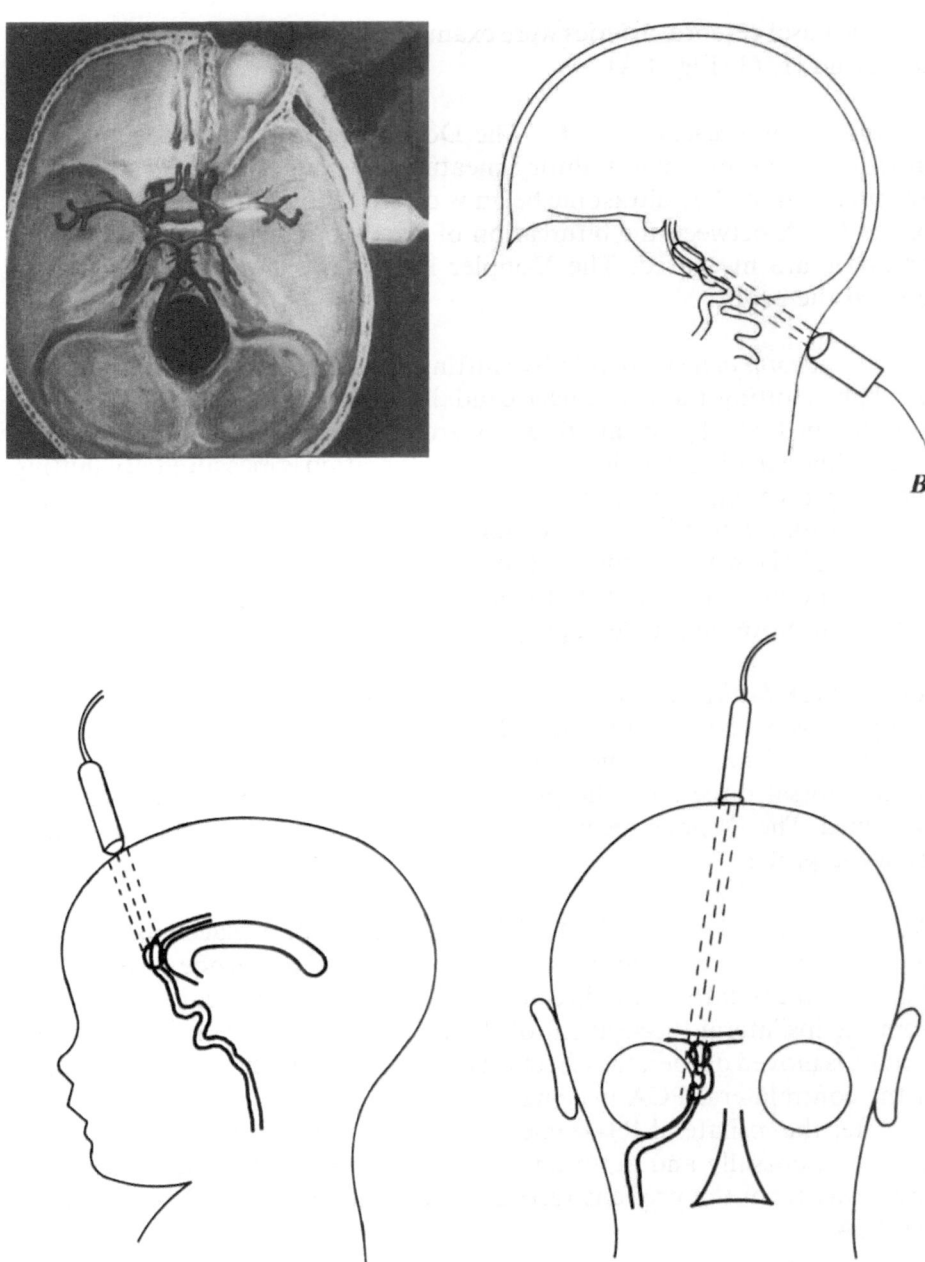

Fig. 4. Recording technique, **A** transtemporal (examination of MCA, ICA, ACA, PCA). **B** Suboccipital through the foramen magnum (examination of the BA). **C** Transcranial high-frontally (examination of the ACA in infants). **D** Through the anterior fontanelle (examination of the ICA in infants)

The basal cerebral arteries were examined using the technique described by Aaslid [1, 6] (Fig. 4 A).

Middle cerebral artery (MCA): The Doppler probe is placed about 1 cm in front of the external auditory meatus and roughly 1–2 cm above the zygomatic arch. The ultrasonic beam is directed horizontally. The segment of the MCA between the bifurcation of the carotid artery and the insular branches are measured. The Doppler recordings show that the flow is toward the probe.

Internal carotid artery (ICA): By shifting the sample volume medially and by slightly tilting the transducer caudally and forwards, Doppler signals are obtained which indicate flow toward the probe. The intracranial segment of the carotid artery just before the bifurcation is measured. By shifting the sample volume further toward the midline and tilting the transducer ventrally and caudally, a bidirectional flow is recorded from the carotid siphon (SIPH), which suddenly stops as the sample volume is shifted further toward the midline. The SIPH can also be insonated transorbitally at a somewhat more favorable angle [360].

Anterior cerebral artery (ACA): After the MCA has been recorded the Doppler probe is tilted forwards and the sample volume is moved toward the midline. The ultrasonic beam is directed horizontally or in a slight cranial-dorsal direction. The precommunicating segment of the ACA is examined. The Doppler recordings reveal that the direction of flow is away from the probe.

Posterior cerebral artery (PCA): The Doppler probe is placed just in front of—or directly over—the helix of the ear. When the probe is tilted just slightly in a caudal/ventral direction, a flow towards the probe originating from the ipsilateral P 1-segment of the PCA is detected. When the sample volume is moved deeper, the direction of flow reverses when the P 1-segment of the contralateral PCA is being insonated.

After the ipsilateral P 1-segment has been recorded, the probe is tilted somewhat dorsally and at an unchanged or only slightly altered depth, a flow away from the probe is recorded from the ipsilateral P 2-segment of the PCA.

Basilar artery (BA): The Doppler probe is placed in the midline between the occiput and the highest vertebra that can be felt. The ultrasound beam is directed cranially and ventrally through the foramen magnum towards the bregma (Fig. 4B). The intracranial segments of the vertebral arteries are first insonated. When the sample volume is moved deeper, the flow in the BA (also away from the probe) is found, which then suddenly stops. The maximum Doppler signal just proximal from here is evaluated.

It is known from anatomical [219] and angiographic [91, 215, 287, 393] examinations that under this technique the angle between the vessel and the ultrasonic beam is well under 30 degrees for the MCA, ACA, and BA, for the ICA, SIPH, and PCA (P 1 and P 2 segments) there can be somewhat larger angles. More recent studies [105a] using the 3-dimensional transcranial Doppler scanner developed by Aaslid [6], which are more applicable to this methodology, demonstrate that the angles of insonation for the ACA and MCA do not exceed 20°, even in contralateral vessels.

In young children, the MCA more so than the ACA ascends higher and steeper than in adults [91, 344, 394]. This must be kept in mind when directing the ultrasonic beam. As for the vertebral and basilar arteries, there are no such differences between children and adults [393].

The MCA, ICA, and ACA in infants are in close proximity to one another. The sample volume of the instrument used is too large to reliably differentiate these vessels using the described examination techniques. The following techniques have therefore been developed to examine the ICA and ACA in infants:

ACA: The Doppler probe is placed high-frontally in the midline with the ultrasonic beam directed caudally and slightly backwards through the thin frontal bone. The flow toward the Doppler probe from the ascending postcommunicating segment of the ACA is measured at an acute angle (Fig. 4C). The vessels of each side are so closely adjacent that the signals from them cannot be differentiated.

ICA: The Doppler probe is positioned on the anterior fontanelle at the midline. The ultrasonic beam is directed caudally and slightly laterally. Flow in the ICA towards the probe is measured just above the cranial base (Fig. 4D). For orientation purposes, the jugular vein is located, which can be identified by continuous flow away from the probe. The ICA is insonated at an acute angle since—unlike in adults—there is very little or no kinking [215, 344].

MCA: The MCA in infants is examined as far laterally as possible according to the technique described by Aaslid [1, 6].

In infants, the PCA and BA are not routinely recorded. They are difficult to find and rarely provide additional information.

An additional help in identifying the basal arteries of the brain is the compression test [1, 6, 160–162]. This test is especially used to identify the precommunicating segment of the ACA. When the ipsilateral common carotid artery is compressed, the flow velocity in the ipsilateral ACA de-

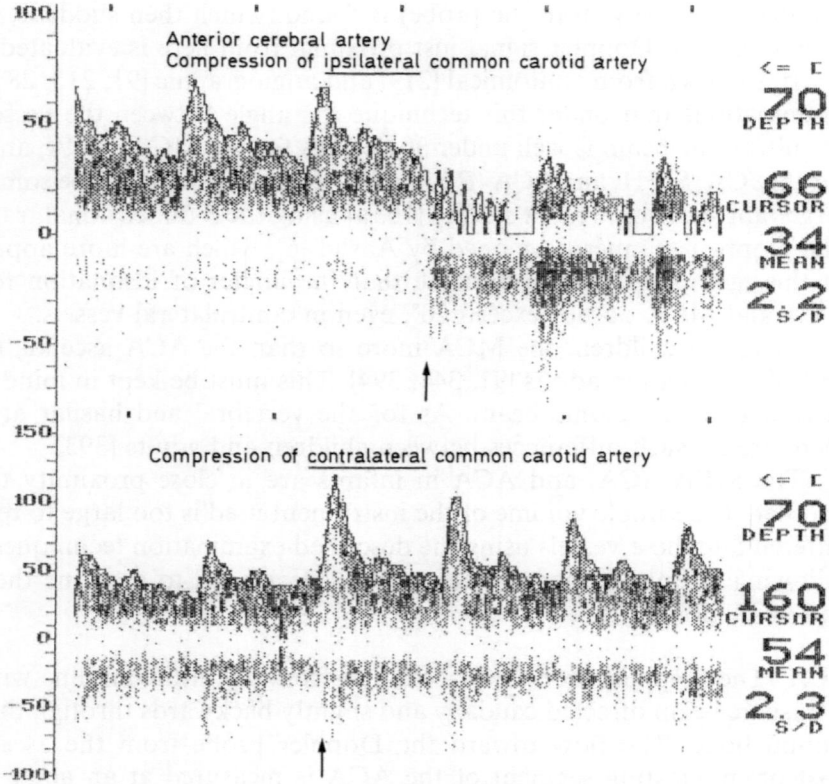

Fig. 5. Compression test. Doppler signals from the ipsilateral ACA (precommunicating segment) when ipsilateral and contralateral common carotid artery are compressed

creases considerably. The waveform of the Doppler spectrum is damped and the direction of flow can even be reversed. Compression of the contralateral common carotid artery produces a marked increase in flow velocity (Fig. 5). These phenomena demonstrate the collateral function of the ACA in the anterior segment of the circle of Willis.

3.4. Procedure

The following parameters of the Doppler spectrum were evaluated in the examinations presented (Chapter 2.4.4.):
— systolic peak blood flow velocity (vs)
— mean peak flow velocity (MEAN = vm)
— enddiastolic peak flow velocity (vd)
— SD ratio and resistance index RI

An average of 3 cerebral arteries per patient were examined according to the method described in Chapter 3.2.

Unless otherwise stated, measurements in infants were carried out during NREM sleep, which was defined by behavioral observations [59, 188, 313, 398]. In most cases this was done at least one hour after the last meal. Examinations in older children were performed while they were awake. Infants were examined in their spontaneous head and body positions, older children on their back with their head being in the midline. The BA was recorded with the children lying on their side.

The measurements were documented and evaluated only when the signals had remained constant for at least 15 seconds.

As a rule, children with pathological Doppler findings were examined several times. To some extent, additional diagnostic examinations were performed including sonography and x-ray of the skull, measurement of the fontanelle pressure, electroencephalogram, computed tomography, and angiography. The results of these examinations will be mentioned only if they are relevant for the evaluation of the Doppler findings.

The results were obtained in systematic studies and from case analyses. In the following, the examination approach is presented together with the respective results. The design of four individual studies is presented beforehand:

1. Fifty healthy preterm and term newborns [288] aged 1 to 10 days were examined. The group had the following characteristics (mean and standard deviation): birth weight (g) $2,503 \pm 726$; current weight (g) $2,422 \pm 700$; length (cm) 45.9 ± 4.7; age (days) 5.3 ± 2.9; gestational age (weeks) 36.5 ± 2.7; capillary hematocrit (%) 56.7 ± 7.1; total bilirubin (mg/dl) 9.1 ± 2.9. Nineteen variables were determined for each child.

2. Forty healthy preterm infants [288] aged between 11 and 69 days were examined. The group had the following characteristics (mean $\pm$ standard deviation): birth weight (g) $1,653 \pm 363$; current weight (g) $2,008 \pm 315$; length (cm) 43.7 ± 2.2; age (days) 30.1 ± 14.3; gestational age (weeks) 33.1 ± 2.2; capillary hematocrit (%) 43.2 ± 9.3.

3. Twenty-four healthy preterm and term newborns (birth weight $2,248 \pm 776$ g, age of gestation (35.6 ± 3.0 weeks) were examined as far as possible on days 1, 2, 3, 4, 7, 10, 15, 20, and 25 (Appendix V). The flow velocities were measured while the babies were awake. The systolic, mean, and diastolic blood pressure as well as the heart rate were measured simultaneously during each measurement by oscillometry (Dinamap 847). In addition, current weight and capillary hematocrit were recorded. This produced a total of 26 variables for each child.

4. One hundred and twelve healthy children aged between 1 day and 16 11/12 years were examined. Age distribution was as follows (number

of children): 1–10 days (18); 11–90 days (14); 3–11.9 months (13); 1–2.9 years (9); 3–5.9 years (18); 6–9.9 years (20); 10–16.9 years (20).

B-mode cerebral sonography (ATL Mark 100) performed in all of the children up to 1 year of age yielded normal findings.

In children under 12 months the MCA, ICA, and ACA were examined and in older children the MCA, ICA, SIPH, ACA, PCA 1, PCA 2, and BA. In each child the vs, vm, vd, SD, and RI were determined. The approximately 8,000 single recordings obtained in these examinations were stored on disks and evaluated using an Olivetti M 24 personal computer.

4. Results
4.1. Methods

4.1.1. Comparison of methods: duplex scan sonography versus transcranial Doppler sonography

Six preterm and term newborns were examined. Each baby was first examined by duplex scan sonography DSS (ATL Mark 500) according to the method described by Jorch [191, 196]. Immediately after, the children were examined by transcranial Doppler sonography TCD. The vs, vm, and vd in the ICA and ACA were measured. Optimal Doppler signals in the ICA were obtained with both techniques at the same depth. The optimal measuring depth for the ACA was somewhat greater with the TCD measurements.

The two methods recorded similar flow velocities in the ICA, more so than in the ACA. TCD found somewhat higher flow velocities in the ICA and especially in the ACA (Figs. 6A and B).

DSS was able to measure velocities under 6 cm/sec, which occurred in 3 out of 6 vd's. With TCD they could only be estimated.

4.1.2. Measuring depth and identification of the vessels

The bitemporal diameter of the skull increases most rapidly during the first two years of life (Fig. 7).

The midline of the skull measured from the temporal bone is about 3 cm in the premature newborn and between 4 and 5 cm during infancy. Not until the age of 6 years does the diameter of the skull reach adult size of 6–7 cm. Accordingly, we found cerebral arteries in children to be age-related at different measuring depths than in adults [1, 6, 21, 160, 162]. This is shown in Table 2.

In infants, with the exception of those with pronounced intracranial malformations, it was always possible to reliably identify the MCA, ICA, and ACA at these depths. In children between the ages of 1 and 3 years, a lack of cooperation made it impossible to measure more than 3 to 4 of the 7 vessels. After the age of 3 years, the MCA, ICA, PCA 2, and BA could be recorded in 100%, the ACA in 98%, the PCA 1 in 93%, and the SIPH in 88% of the children.

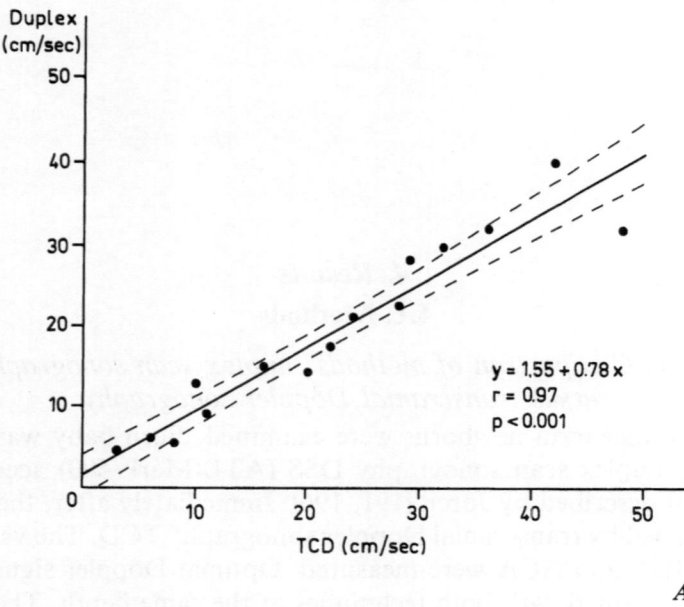

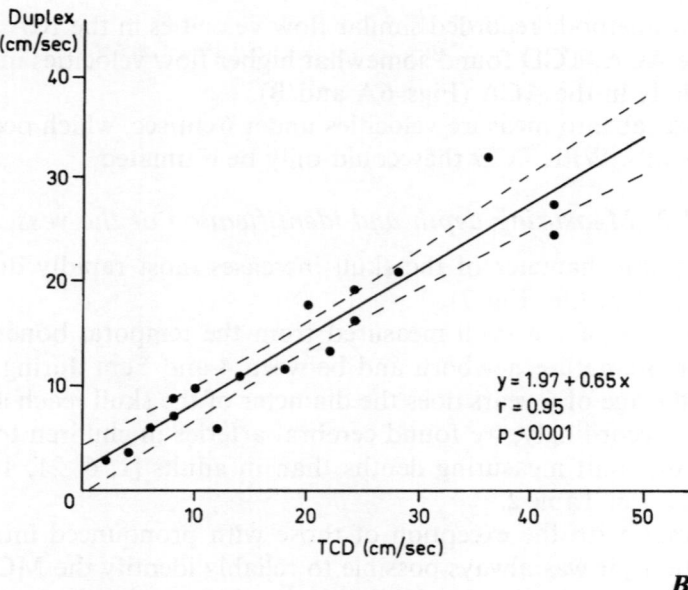

Fig. 6. Comparison of the result recorded by transcranial Doppler sonography (TCD-x) and duplex scan sonography (Duplex-y) in 6 preterm and term newborns. ——— Regression line, − − − 95% confidence interval. **A** Flow velocities in the ICA, **B** flow velocities in the ACA

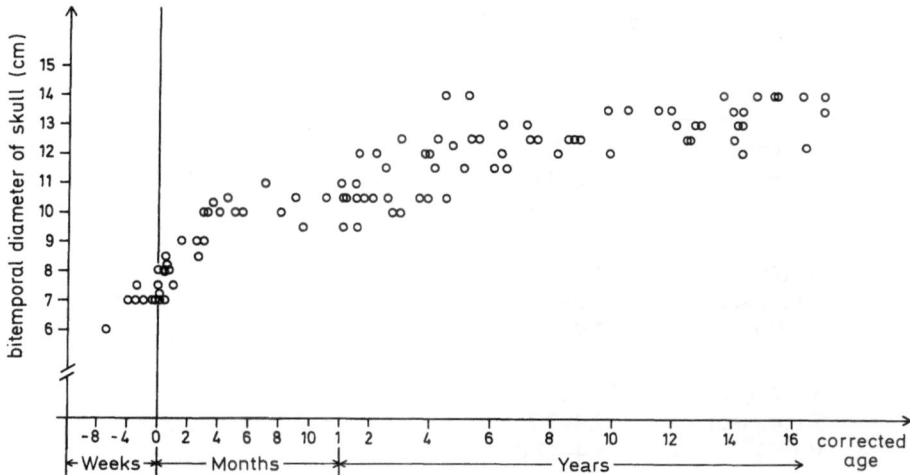

Fig. 7. Bitemporal diameter of the head of 100 randomly selected patients at the University Children's Hospital, Freiburg

Table 2. Depths (in mm) at which the basal cerebral arteries in children were measured

Age	MCA	ICA	SIPH	ACA	PCA 1	PCA 2	BA
0–3 months	25	55–65[a]	—	25–30[b]	—	—	—
3–12 months	30	60–70[a]	—	30[b]	—	—	—
1–3 years	35–45	40–50	50–60	55–65	55	50–55	50–60
3–6 years	40–45	45–55	55–60	60–65	55–60	50–60	55–70
6–10 years	45–50	50–55	55–60	60–70	60–70	55–65	55–75
10–18 years	45–50	55	60	65–70	60–70	60–65	60–80

[a] Examination through the anterior fontanelle
[b] Examination from a high frontal position

4.1.3. Intraindividual variability

Reproducibility

In five infants the vs and the vm in the MCA, ICA, and ACA were recorded a total of 7 times, each at 5-minute intervals. Table 3 shows the reproducibility of the Doppler parameters.

The variation coefficients were somewhat smaller in the MCA than in the ACA and the ICA. The variation coefficients increased in the order RI-vs-vm-SD-vd. All of the Doppler parameters were easily reproducable in all of the vessels examined.

Table 3. Reproducibility of the Doppler parameters
[highest and lowest individual mean value (M), standard deviation (S),
and variation coefficients (V, V = S * 100/M)]

		MCA	ICA	ACA
	M (cm/sec)	61–74	54–86	46–59
vs	S (cm/sec)	1.0–3.0	1.4–4.3	1.2–2.1
	V (%)	1.6–4.1	2.1–7.8	2.2–4.6
	M (cm/sec)	33–39	31–41	25–33
vm	S (cm/sec)	1.0–2.6	0.8–2.6	1.1–1.9
	V (%)	2.9–6.5	2.2–8.1	3.2–6.5
	M (cm/sec)	10–23	15–17	10–19
vd	S (cm/sec)	0.4–2.2	0.5–1.6	0.5–1.6
	V (%)	3.4–12.0	3.1–9.3	4.1–11.2
	M	3.0–7.7	3.3–5.1	2.9–5.4
SD	S	0.1–0.4	0.2–0.3	0.2
	V (%)	2.6–6.6	3.9–6.1	3.7–6.9
	M	0.67–0.87	0.69–0.80	0.66–0.82
RI	S	0.01–0.02	0.01–0.03	0.01–0.02
	V (%)	0.8–2.2	1.2–3.6	0.9–3.4

Comparison of sides

Consecutive measurements of the left and right MCA and ICA performed in 20 preterm and term newborns revealed variation coefficients of 0–3% (vs), 0–4% (vm), 0–6% (vd), 2–5% (SD), and 0–3% (RI). In 20 children aged 6–12 years, measurements of the left and right sides yielded variation coefficients of 0–4% for the MCA and 3–8% for the ACA and PCA. The results show that except for special cases (Chapter 4.4), it is sufficient to examine the vessels on one side.

Correlation between flow velocities

In an analysis of the correlation coefficient matrices [358] listed in Appendices I–III, the following relationships were found:

1. The correlation coefficients for the relations between vs, vm, and vd in the MCA, ICA, and ACA were always so high that a linear relationship between the flow velocities could be assumed. The velocities correlated better in the MCA and ACA than those recorded in the ICA. In

older children the velocities in the MCA and ICA correlated better with one another than each of them did with the velocities recorded in the ACA.

2. The flow velocities found in one vessel correlated better with one another than they did with the velocities measured in other vessels.

3. Of the flow velocities in the various vessels, the vs, vm, and vd correlated best. The correlations between vs and vm were somewhat better than those between vs and vd and vm and vd.

These relationships indicated that particular attention should be paid to the vs and vm measured in the MCA.

4.2. Physiological influences on the Doppler parameters

4.2.1. Position of the head and body

When preterm and term newborns were quickly changed from a horizontal to a vertical position, a marked regression of all flow velocities could be observed in some cases and no change in others. Alterations in vigilance had a substantial effect on the results.

Ten healthy premature newborns were examined, first with the head raised (25° to horizontal) and then with the head lowered (15°). The vs, vm, and vd in the MCA were recorded in each child (Table 4).

The variation coefficients for pairs of measurements (head raised and head lowered position in each subject) were between 2.5% (vs) and 6.3% (vd). Thus, in this examination the position of the body apparently did not exert any influence on the flow velocities.

The flow velocities in the MCA were recorded in five infants in the following positions:

— supine position, head in the midline—MCA left and right
— supine position, head turned to the left—MCA right
— supine position, head turned to the right—MCA left
— lying on the left side—MCA right
— lying on the right side—MCA left

Table 4. Effect of body position on the flow velocities in the MCA in 10 preterm infants (mean value M and standard deviation S)

	Head raised		Head lowered	
	M	S	M	S
vs (cm/sec)	58.6	15.2	59.8	15.7
vm (cm/sec)	28.0	10.3	28.6	10.4
vd (cm/sec)	10.6	5.8	10.8	6.3

— prone position, head turned to the left—MCA left
— prone position, head turned to the right—MCA right

The examination took about 2 hours for each infant. The variation coefficients in each child ranged from 2.1–4.0 (vs), 2.8–6.9 (vm), and 6.4–13.6 (vd).

The position of the body did not have any marked effect on the flow velocities, nor was there any difference between sides. Over a period of 2 hours in which the infants remained in the same state of vigilance, the flow velocities were virtually constant.

4.2.2. Vigilance

The flow velocities in preterm newborns, term newborns, and young infants were strongly dependent on the state of vigilance (Fig. 8). Velocities remained constant in NREM sleep. We evaluated the measurements only if the velocities remained constant for at least 15 seconds.

In children older than 1 year we occasionally found velocities to be higher at the beginning of the examination than after the child had become accustomed to the situation and quiet. For examinations on children in this age group an acclimation phase lasting a few minutes preceeded the recording phase.

4.2.3. Age

The vs, vm, vd, SD, and RI were clearly age-dependent in all of the cerebral vessels examined. The flow velocities increased linearly during the first 2 months of life. Up to the age of 5 to 6 years they increased more slowly to approximately 4 times the initial value measured at birth. Up to the age of 16 years they decreased linearly until they reached approximately 70% of the maximum value. Fig. 9 shows an example.

The flow velocities decreased in the order MCA-ICA-SIPH-ACA-BA-PCA 1-PCA 2.

SD and RI decreased only during the first year. They presented almost identical values in all of the vessels examined.

The age-dependent values measured for vs, vm, vd, SD, and RI in 7 basal arteries are listed in Appendices IVa to IVe. The interindividual variation coefficients of the Doppler parameters can be calculated from the mean values and standard deviations given.

During the first weeks, of life the flow velocities increased most rapidly. During this time the following relationships were found:

1. The relationships between age and flow velocities in the MCA, ICA, and ACA were described by linear relationships (Figs. 10A and 10B). The vs showed the closest correlation with age, followed by vm and then vd. The average velocity increase per day for vs, vm, and vd was 1.5, 0.8, and

Preterm newborn, 35th week of gestation, 2100 g, 5th day of life

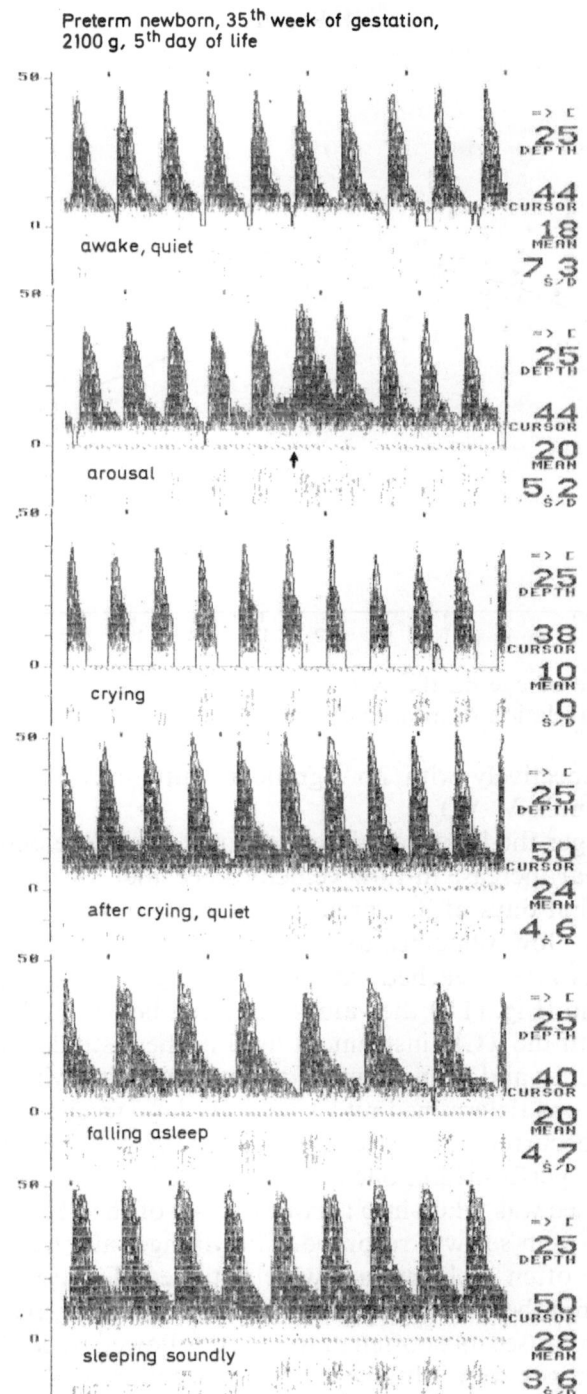

Fig. 8. Influence of vigilance and behavior on the Doppler signals from the MCA in a 3 week old premature infant born in the 32nd week of gestation

4. Results

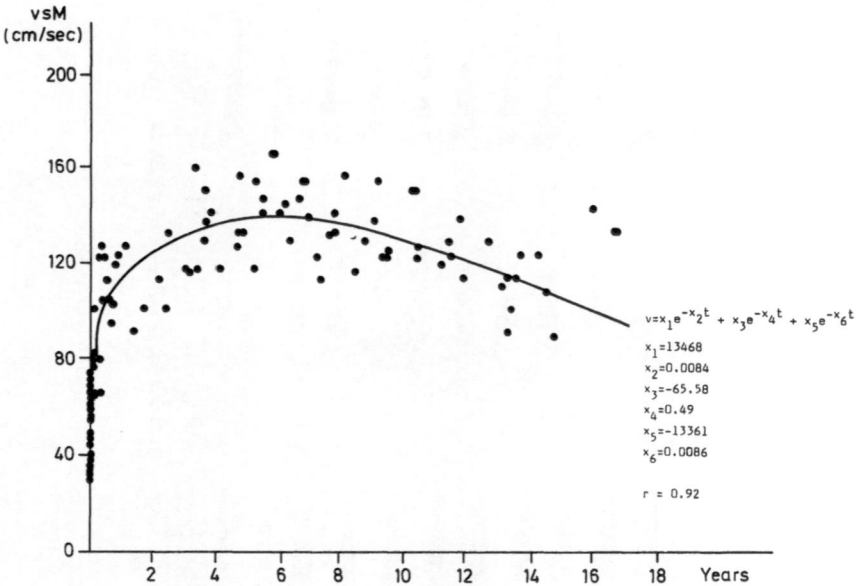

Fig. 9. Influence of age on the systolic peak flow velocity in the MCA (vsM). (*t* Age, *r* correlation of measured values and estimated values) [285, 361]

0.4 cm/sec, respectively with no significant difference for various birth weights (Appendix V, VI).

2. The longer the baby could be examined, the more definite the age-dependent linear increase in flow velocities became.

3. In the first days of life, preterm and term newborns often showed a return of the flow velocities to very low values after the systolic peak flow velocity had been reached. At the beginning of diastole (Fig. 11 A) or towards the end (Fig. 11 B) the values often fell below the lowest velocity measurable with the TCD instrument used in these studies (6 cm/sec).

In 155 healthy and ill preterm and term newborns within the first 30 days of life, it was determined retrospectively how frequently a vd of less than 6 cm/sec occurred in the MCA, ICA, and ACA. When multiple measurements were performed in one child, only the last time the vd had been less than 6 cm/sec was taken into account. In 40 of the 155 children (26%) vd of less than 6 cm/sec was recorded. The younger and lighter the babies were, the more often such findings were obtained (Table 5).

In 14 children the vd was below 6 cm/sec in the MCA, ICA, and ACA. The vd was < 6 cm/sec most often in the ICA, followed by MCA and ACA. Similar percentages were obtained when the children were grouped according to gestational age instead of birth weight. Also, eliminating the eight children with patent ductus arteriosus Botalli did not greatly alter the percentages.

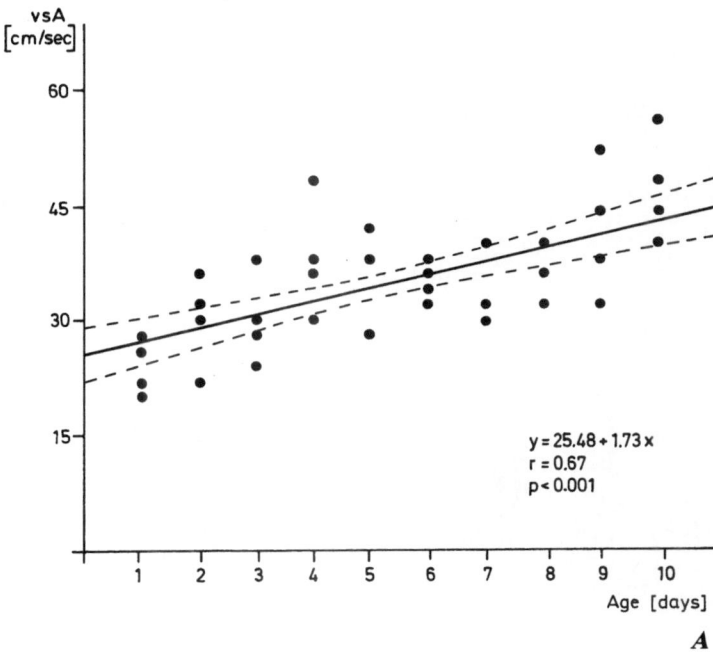

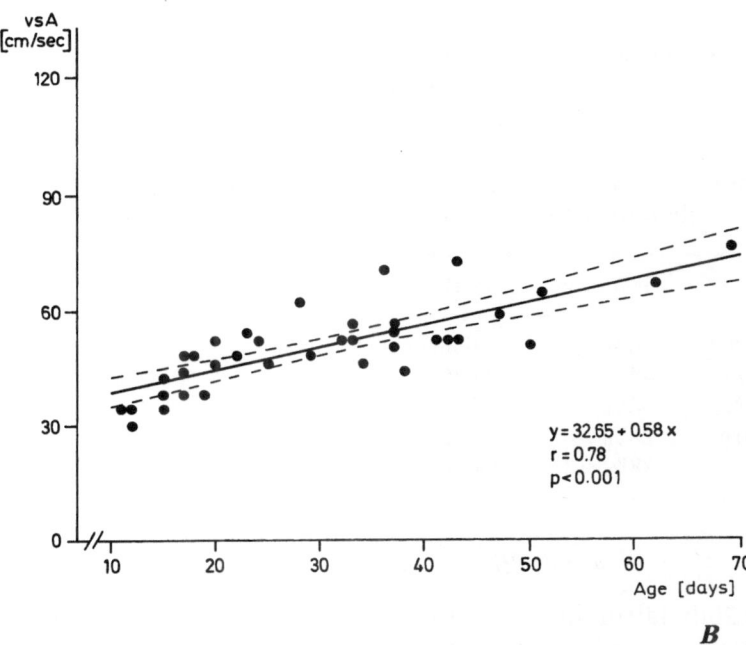

Fig. 10. Influence of age on the flow velocities in the first weeks of life.
——— Regression line, – – – 95% confidence interval. **A** In the first 10 days,
B after the 10th day (*vsA* systolic peak flow velocity in the ACA)

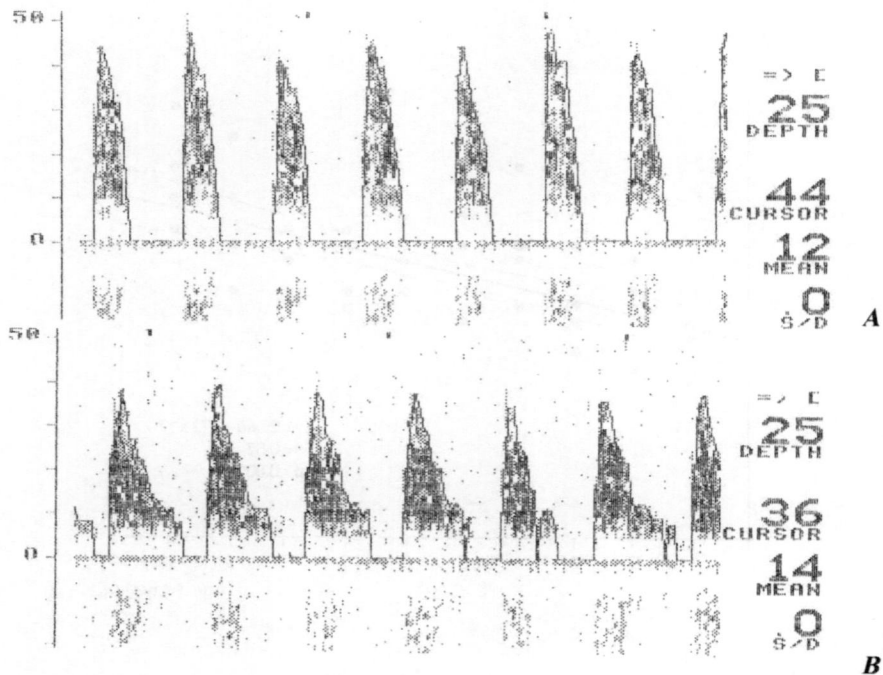

Fig. 11. Low enddiastolic flow velocities in the first days of life—two typical examples

Table 5. Percentage of newborns with enddiastolic peak flow velocity (vd) of less than 6 cm/sec—dependence on birth weight and age

	1–5 days	6–10 days	11–20 days	21–30 days	Total
Up to 1,500 g	80%	70%	33%	38%	51%
1,501–2,000 g	44%	56%	23%	0%	29%
2,001–2,500 g	30%	18%	0%	0%	16%
Over 2,500 g	28%	0%	0%	0%	10%
Total	38%	30%	19%	10%	26%

4.2.4. Birth weight/gestational age

Up to the 20th day of life, the relationship between birth weight/gestational age and flow velocities in the MCA, ICA, and ACA could be described by linear relationships which became less definitive in the order vd-vm-vs (Fig. 12, Appendix VI). Like vd, SD and RI correlated with birth weight and gestational age, though negatively.

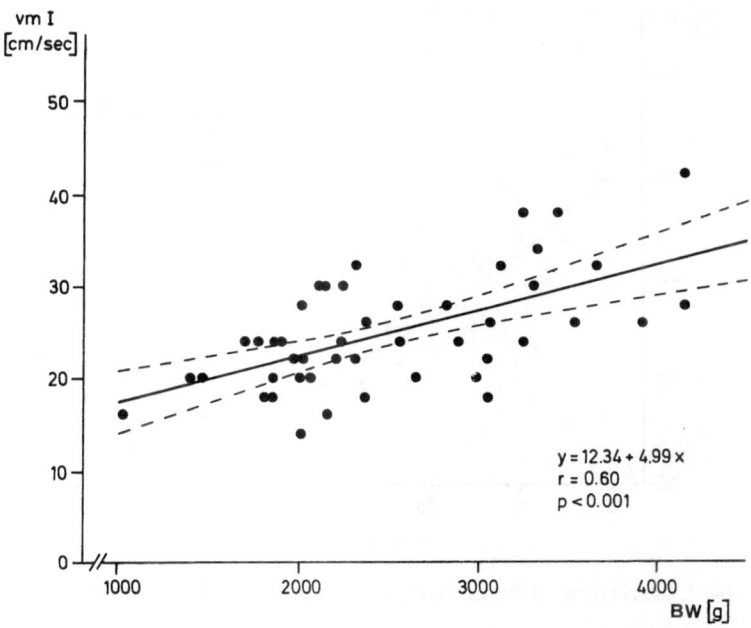

Fig. 12. Influence of birth weight (*BW*) on the flow velocities in the first 10 days of life; *vmI* mean peak flow velocity in the ICA.
———— Regression line, — — — 95% confidence interval

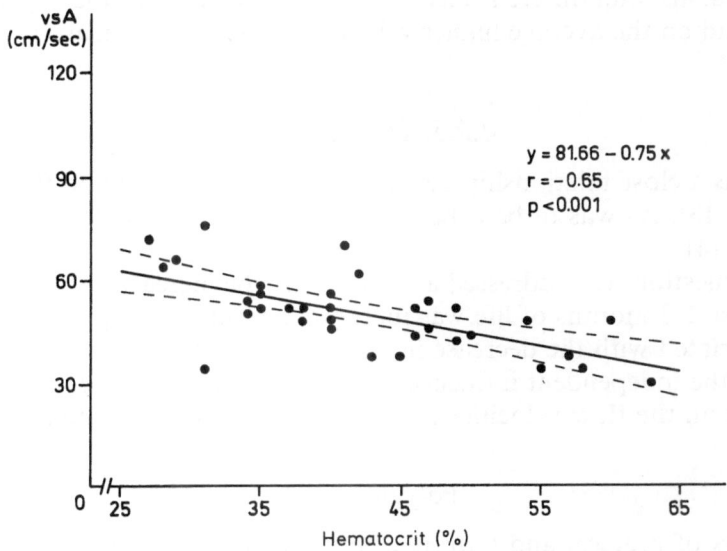

Fig. 13. Correlation of hematocrit and blood flow velocities in healthy premature newborns. (*vsA* Systolic peak flow velocities in the ACA.)
———— Regression line, — — — 95% confidence interval

4. Results

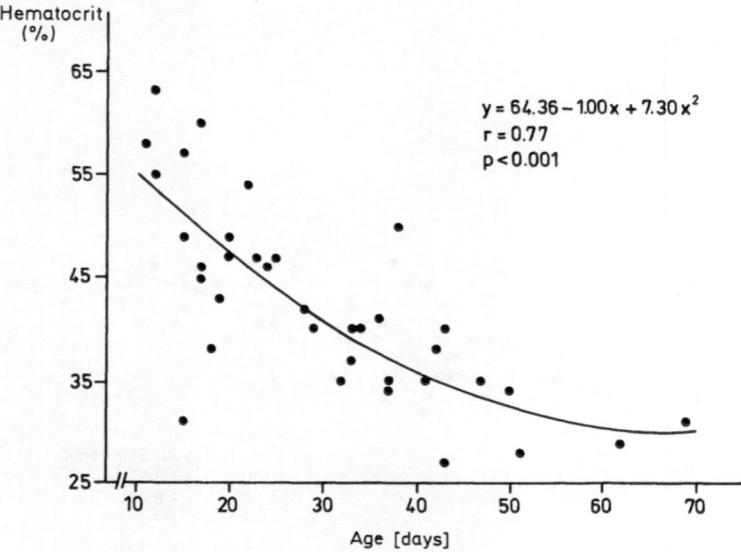

Fig. 14. Correlation of hematocrit and age in healthy premature infants

Children with lower birth weight/gestational age had lower vd values more frequently and for a longer time (Table 5). In children of the same gestational age and different birth weights, the child with the higher birth weight had on the average higher velocities. This was particularly true for twins.

4.2.5. Hematocrit

There was a close relationship between hematocrit (HCT) and flow velocities (Fig. 13). As was to be expected, the hematocrit also correlated with age (Fig. 14).

The question was addressed as to whether the increase in flow velocity in the first 2–3 months of life was merely the result of aging or whether it was associated with the decrease in hematocrit with age. Using pair comparisons the independent influence of a HCT that strongly deviated from the norm on the flow velocities in the MCA was demonstrated:

Polyglobulia

Nine pairs of preterm and term newborns with normal or elevated HCT relative to age [284] were compared (Table 6).

Children with elevated HCT had significantly lower vs than those with normal HCT.

Table 6. Flow velocities in the MCA in the presence of normal and elevated hematocrit—pair comparison (mean values)

	Normal HCT	Elevated HCT	t-Test
Age (days)	3.0	2.4	n.s.
Gestational age (weeks)	37.0	37.5	n.s.
Birth weight (g)	2,707	2,605	n.s.
HCT (%)	52.6	72.0	$p < 0.0001$
vs (cm/sec)	52	40	$p < 0.01$
vm (cm/sec)	24	20	n.s.
vd (cm/sec)	14	11	n.s.
RI	0.70	0.73	n.s.

Table 7. Flow velocities in the MCA in the presence of normal and lowered HCT—pair comparison (mean values)

	Normal HCT	Reduced HCT	t-Test
Age (days)	21	21	n.s.
Gestational age (weeks)	35.4	34.3	n.s.
Birth weight (g)	2,030	1,999	n.s.
HCT (%)	51	33	$p < 0.001$
vs (cm/sec)	55	68	$p < 0.06$
vm (cm/sec)	28	37	$p < 0.05$
vd (cm/sec)	14	22	$p < 0.05$
RI	0.75	0.68	$p < 0.05$

Anemia

Eight pairs of preterm newborns with normal and lowered HCT [284] were compared (Table 7).

Children with lowered HCT had significantly higher values for vs, vm, vd, and RI.

4.2.6. Arterial blood pressure

The mean values of flow velocities and blood pressure measured on defined days of life during the examination of the MCA, ICA and ACA showed an almost parallel increase. The value of the systolic blood pressure was just above that of the vs in the MCA, the values of the mean and diastolic blood pressure were about twice that of vm and vd in the MCA (Fig. 15).

There were no significant correlations, however, when the 4,500 flow velocities were compared with the blood pressures measured at the same

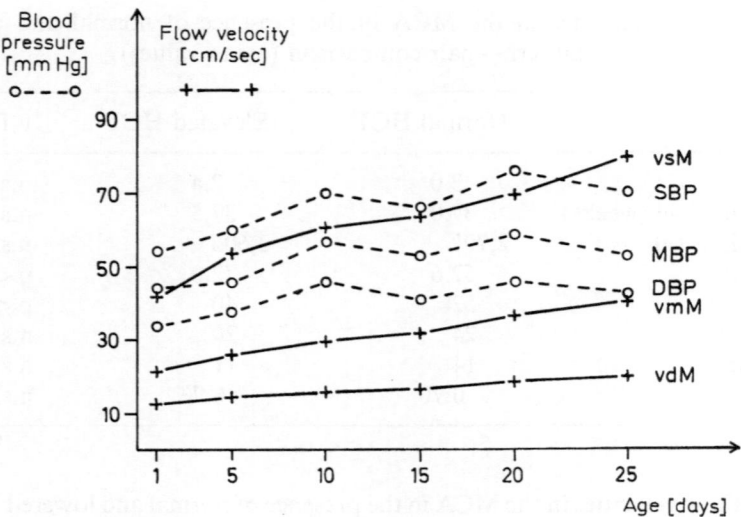

Fig. 15. Mean values of blood pressures and flow velocities in 40 healthy premature newborns showing dependence on age: *vsM* systolic, *vmM* mean, *vdM* enddiastolic, peak flow velocity in the MCA, *SBP* systolic, *MBP* mean, *DBP* diastolic blood pressure

time. When vs, vm, and vd were correlated with the respective blood pressures in each child, the correlation was significant only in 25 of the 216 possible cases (chance effect).

4.2.7. Heart rate

In the longitudinal study on healthy preterm and term newborns the heart rate of the children examined was always above 120/min. A significant relationship between the vs, vm, and vd with the respective heart rate was only established in 7 of the 216 possible cases (chance effect).

In infants with bradycardia a marked reduction of vd, followed by a decrease of vs was observed (Fig. 16).

In older children with bradycardia no change in the relationship of the flow velocities with one another was observed. The velocities in diastole dropped more slowly in accordance with the heart rate (Fig. 17).

Considerable abnormalities occurred in the presence of cardiac arrhythmia (Fig. 18).

4.2.8. Bilirubin and phototherapy

Severe hyperbilirubinemia

In five term newborns with a total serum bilirubin level of over 20 mg/dl the flow velocities were normal for age and weight.

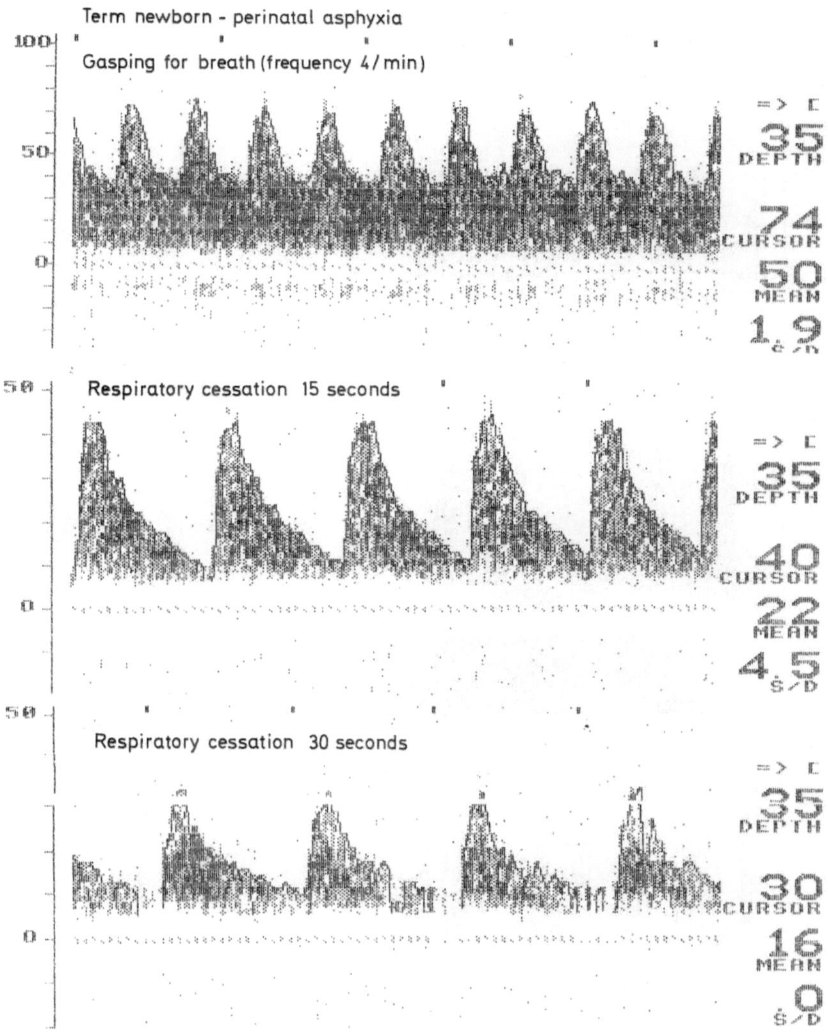

Term newborn - perinatal asphyxia

Gasping for breath (frequency 4/min)

```
=> [
35
DEPTH

74
CURSOR

50
MEAN

1.9
s/D
```

Respiratory cessation 15 seconds

```
=> [
35
DEPTH

40
CURSOR

22
MEAN

4.5
s/D
```

Respiratory cessation 30 seconds

```
=> [
35
DEPTH

30
CURSOR

16
MEAN

0
s/D
```

Fig. 16. Influence of bradycardia on the Doppler signals obtained from the MCA—newborn after perinatal asphyxia

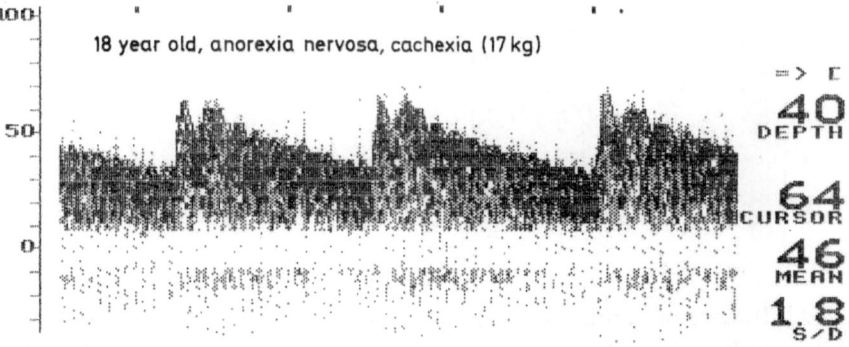

18 year old, anorexia nervosa, cachexia (17 kg)

```
=> [
40
DEPTH

64
CURSOR

46
MEAN

1.8
s/D
```

Fig. 17. Influence of bradycardia (heart rate 36/min) on the Doppler signals obtained from the MCA in a youth

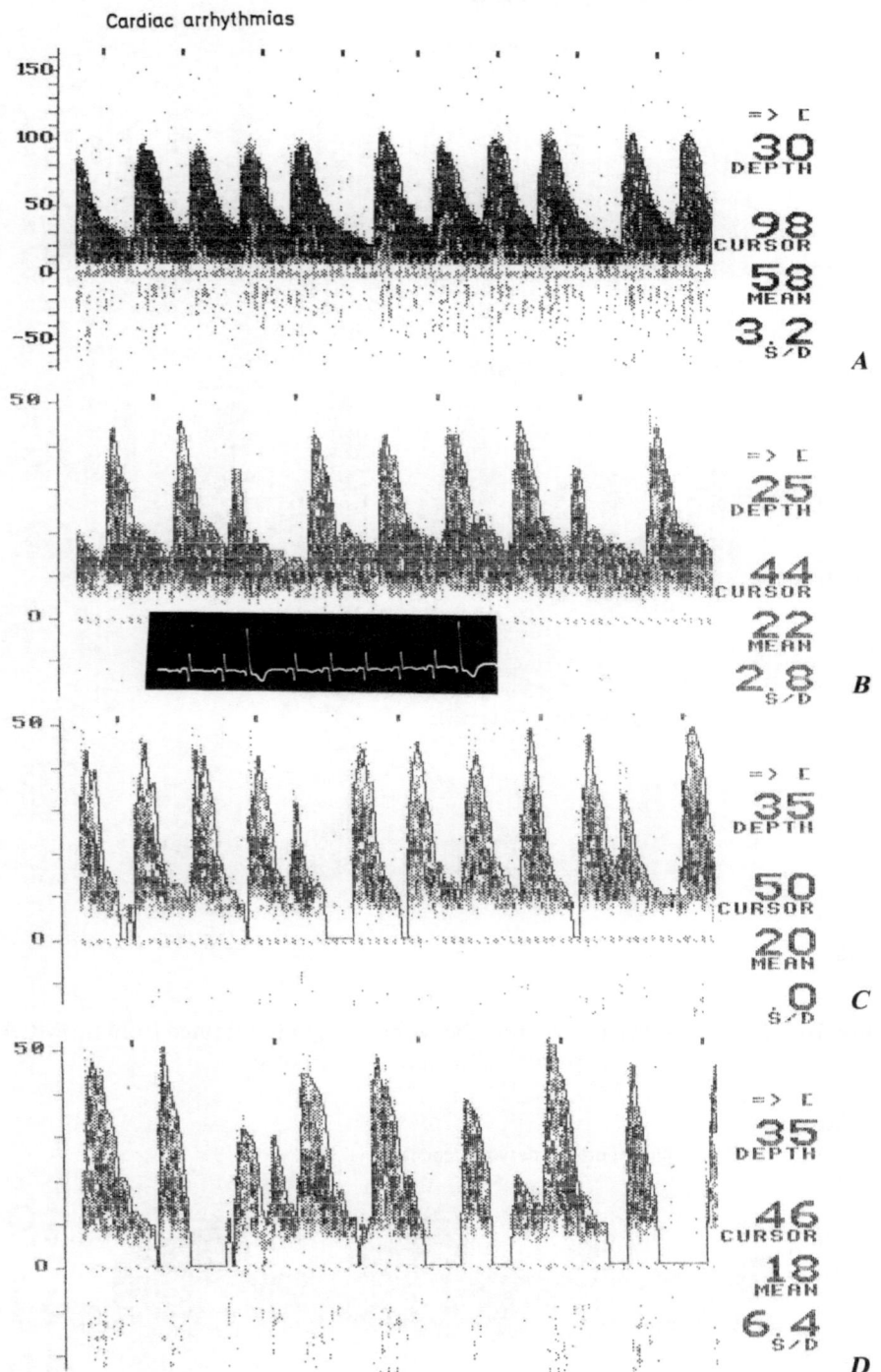

Fig. 18. Influence of cardiac arrhythmias on the Doppler signals obtained from the MCA in four infants (in B, electrocardiogram is shown on the black background)

Phototherapy

Since phototherapy is carried out only when the bilirubin values ar elevated [413], it was not possible to examine the effect of bilirubin and phototherapy on the flow velocities separately.

1. Ten pairs of preterm and term newborns similar in age, gestational age, and birth weight were examined. One group had been receiving continuous phototherapy and intravenous fluid due to hyperbilirubinemia for 12 hours prior to the examination; the other group without hyperbilirubinemia had not been treated before the measurements.

The phototherapy did not have a significant influence on the flow velocities.

2. The flow velocities in the MCA were examined in 10 newborns with hyperbilirubinemia. Measurements were performed immediately before phototherapy, 12 hours after the start and 12 hours after discontinuing phototherapy.

The flow velocities increased during that time, but the increase was not significant.

4.2.9. CO_2 partial pressure

The flow velocities in the MCA were determined in 10 preterm and term newborns without sonographic evidence of intracerebral hemorrhage under controlled ventilation before (1) and after (2) a change in the transcutaneously measured CO_2 partial pressure had occurred (Transcapnode, Hellige, Freiburg, Federal Republic of Germany). The measurements took place at intervals of a maximum of one hour (Table 8).

Table 8. Influence of the CO_2 partial pressure (in mm Hg) on the mean peak flow velocity (vm, in cm/sec) in the MCA in ventilated preterm and term newborns

Patient	CO_2 (1)	CO_2 (2)	vm (1)	vm (2)	$\Delta\ CO_2$	Δ vm	Δ vm/$\Delta\ CO_2$
1	21	37	18	56	16	38	2.4
2	32	39	10	20	7	10	1.4
3	38	51	15	48	13	33	2.5
4	36	42	12	18	6	6	1.0
5	51	56	22	32	5	10	2.0
6	68	78	32	50	10	18	1.8
7	65	73	32	42	8	10	1.3
8	61	37	20	8	24	12	0.5
9	57	43	50	25	14	21	1.5
10	49	38	42	24	11	18	1.6
Mean value							1.6

Preterm infant, 31st week of gestation

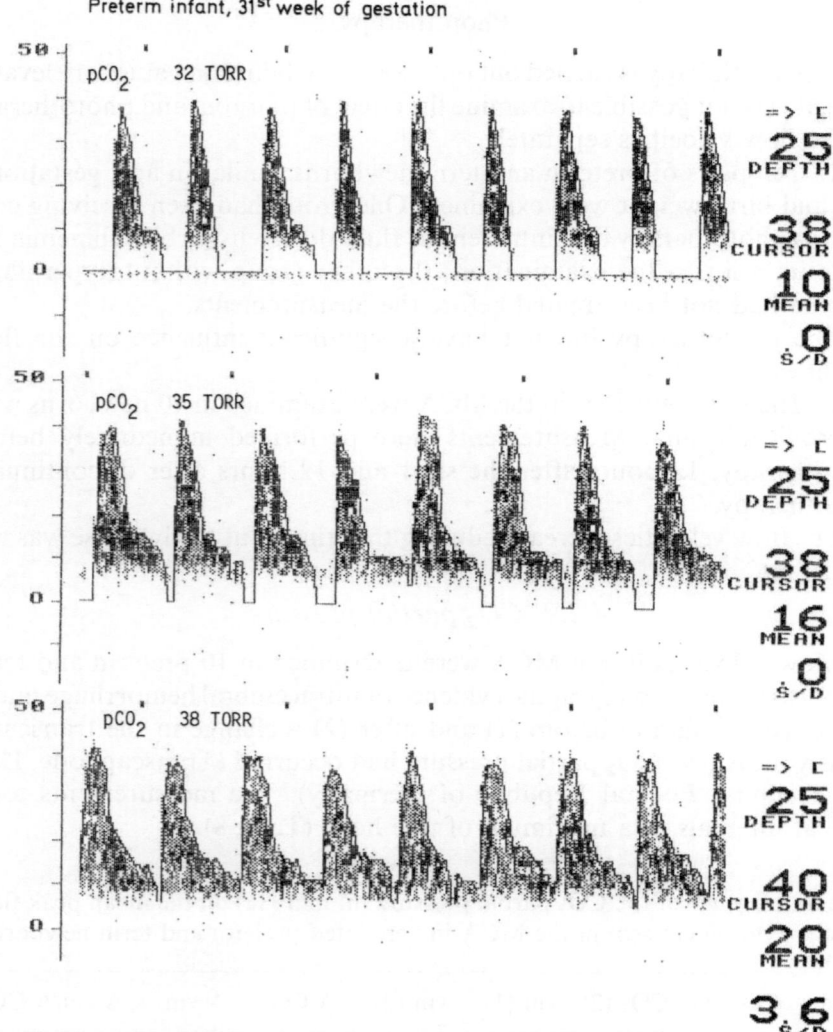

Fig. 19. Dependence of Doppler signals from the MCA on transcutaneous measurement of partial pressure in a premature infant under controlled ventilation

On the average, vm changed by 1.6 cm/sec per mm Hg CO_2-change. The correlation between CO_2 and vm was only minimal (r = 0.42; p = 0.23), owing to the considerable interindividual differences.

In the presence of slight hypocapnia an initial reduction of vd was observed (Fig. 19). In marked hypocapnia there was an additional decrease in the vs (Fig. 20). The vd rose more than the vs in the presence of hypercapnia (Fig. 20). In hypocapnia the SD and RI were higher than the norm; in hypercapnia they were lower.

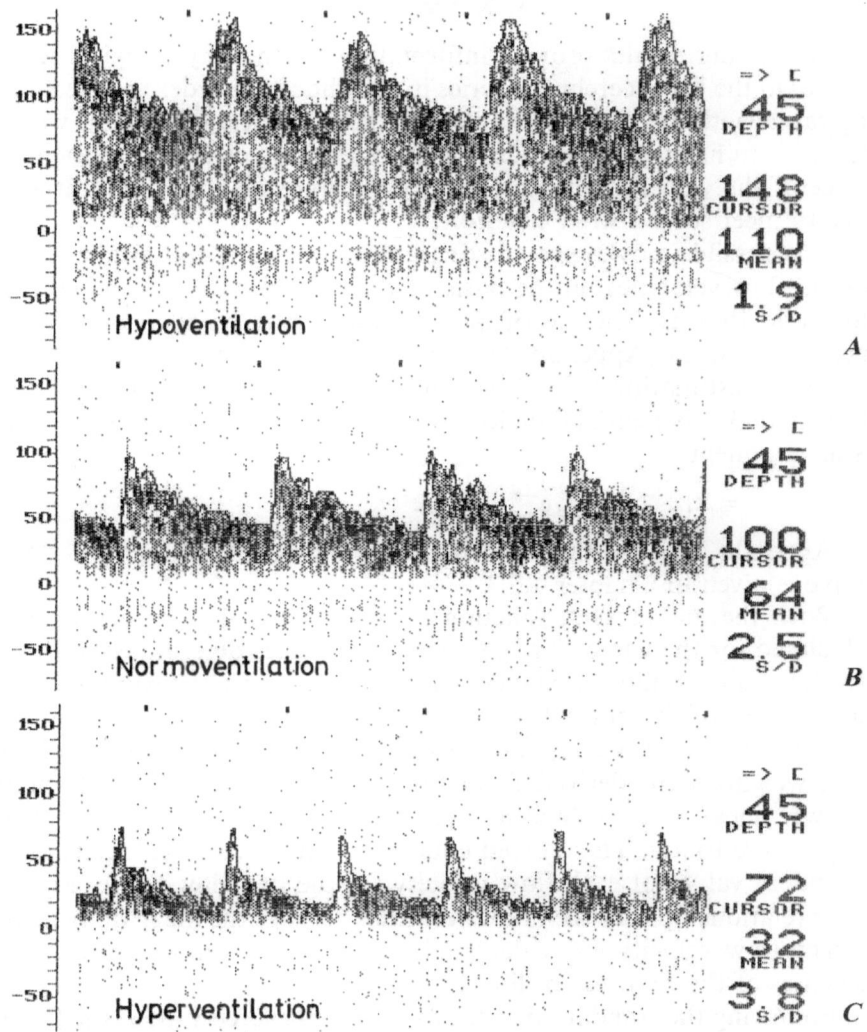

Fig. 20. Changes in the Doppler signals from the MCA of a youth during voluntary hypoventilation and hyperventilation

In the presence of marked cerebral hemorrhage no increase in flow velocities was observed, despite raised CO_2 partial pressures. The CO_2 reactivity of the flow velocities was much lower in children with patent ductus arteriosus and with transcutaneous CO_2 partial pressure of more than 70 mm Hg.

4.3. Reference values

The foregoing results provide information on the physiological flow velocities in the basal cerebral arteries in childhood. In order to detect pathological deviations of the flow velocities in childhood by TCD it would be desirable to have normal values which objectively characterize the normal range of the velocities at each age. The examinations carried out are not able to meet such a strict criterion.

It is possible, however, to distinguish between physiological and pathological flow velocities on the basis of these examinations. The relation— flow velocities as function of age—is quantitatively calculated from the measured values (Appendix IV) according to the method of non-linear parameter estimation [285, 361]. The velocities (y) can be determined by the sum of 3 exponential functions in which the respective age (t) is inserted as an exponent:

$$y = x_1 e^{-x_2 t} + x_3 e^{-x_4 t} + x_5 e^{-x_6 t} \tag{20}$$

Appendix VII shows the constants x_1—x_6 of these functions. An example is given in Chapter 4.2.3.

With the aid of such functions, the expected flow velocities can be calculated for the age of 0 to 18 years. The values determined in this way are designated reference values. They correspond approximately to the mean values of the normal flow velocities (Table 9).

Mean reference values for the first 20 days of life and for different birth weights were derived from the data of the longitudinal study as follows: linear relationships between flow velocities on day 1 and birth weight were found. The derived equations were used to calculate weight-dependent values of each flow velocity for the first day of life. From the linear relationships between flow velocities and age the mean daily increase of each flow velocity was calculated and added to the weight-dependent velocities of the first day of life (Fig. 21).

Following the standard deviations of the measured values, the Doppler flow velocities will be designated "increased" or "decreased" when they deviate more than 30% from the reference values. In addition, comparisons of the two sides of the brain and repeated examinations on the same patient will be considered.

4.4. Clinical applications

4.4.1. Patent ductus arteriosus Botalli

A total of 80 Doppler examinations were conducted in 29 infants with patent ductus arteriosus Botalli (PDA). The PDA was diagnosed according to clinical criteria and echo cardiography [107, 177, 218, 242, 253, 260,

Table 9. Reference values of the flow velocities in the MCA, ICA, and ACA during childhood (in cm/sec)

Age	vsM	vmM	vdM	vsI	vmI	vdI	vsA	vmA	vdA
0–1 months	53	28	17	56	29	16	40	22	14
1–2 months	75	44	25	75	36	21	54	33	18
2–3 months	89	55	32	88	42	25	63	41	22
3–4 months	99	63	37	97	47	28	70	46	25
4–5 months	105	68	41	103	52	32	74	50	28
5–6 months	109	71	45	107	56	35	77	53	31
6–7 months	112	74	48	110	60	37	80	55	33
7–8 months	114	76	50	112	64	40	81	56	34
8–9 months	116	77	52	114	67	42	82	57	36
9–10 months	118	78	53	115	70	45	83	58	37
10–11 months	119	79	55	116	73	47	84	58	38
11–12 months	121	79	56	117	75	48	85	59	39
1–2 years	136	86	64	130	86	65	90	60	43
2–3 years	143	90	66	135	95	68	92	60	44
3–4 years	146	91	68	138	98	69	93	60	44
4–5 years	148	92	69	140	97	68	93	60	44
5–6 years	148	92	69	140	95	67	93	59	44
6–7 years	147	92	69	139	93	66	92	59	43
7–8 years	144	91	68	137	90	64	91	58	43
8–9 years	141	89	67	134	87	62	90	57	43
9–10 years	137	87	66	130	83	61	88	56	42
10–11 years	132	85	65	127	80	59	86	55	42
11–12 years	127	82	63	123	76	57	83	54	41
12–13 years	122	80	61	119	72	54	81	52	40
13–14 years	116	77	59	114	68	52	78	51	39
14–15 years	111	74	58	110	64	50	76	50	39

vs, vm, vd systolic, mean, enddiastolic peak flow velocity
M MCA, *I* ICA, *A* ACA
Thus, vsM means the systolic peak flow velocity in the MCA, etc.

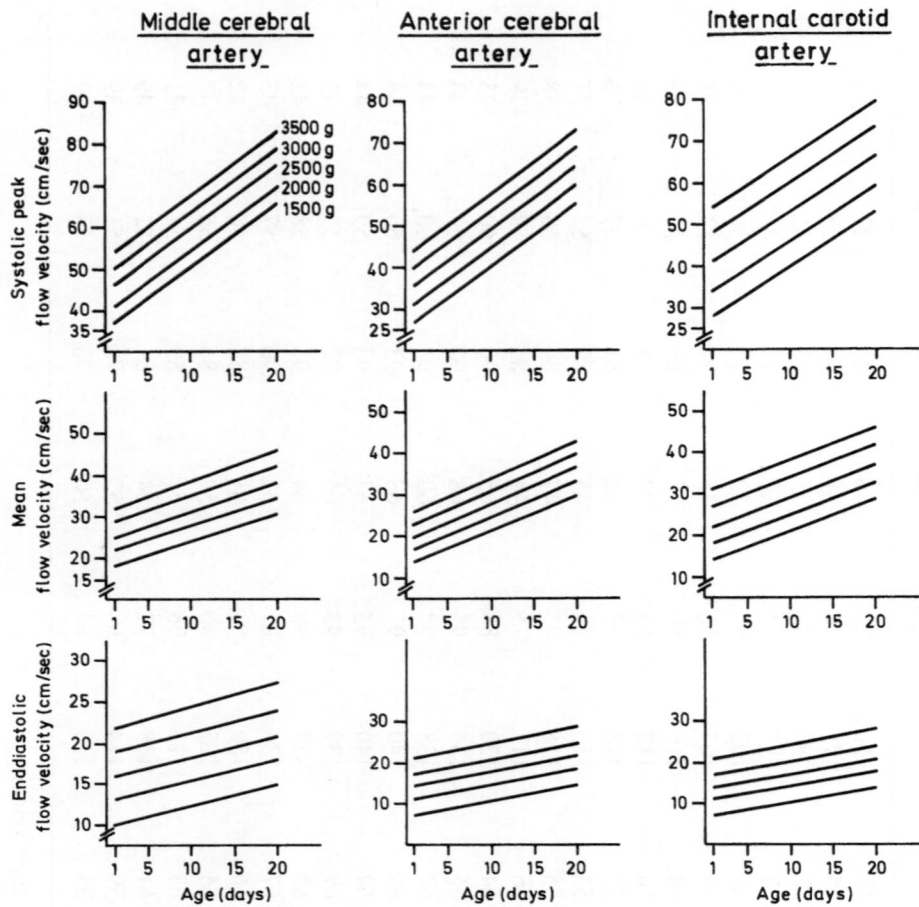

Fig. 21.. Reference values of flow velocities in basal cerebral arteries from 1–20 days of age for different birth weights

301, 334, 392, 414, 415]. Thirteen premature infants had a hemodynamically relevant (=large) PDA, which could be successfully treated (11 × indometacin, 2 × surgical closure). Seven premature newborns had a hemodynamically irrelevant (=small) PDA, which closed spontaneously. In none of the preterms was a cerebral hemorrhage detected in the sonogram. The arterially or transcutaneously measured CO_2 partial pressure was between 40 and 50 mm Hg.

Six infants had a PDA with additional heart defects dependent on the ductus. Three infants had a persistent fetal circulation.

Fig. 22 shows the typical Doppler findings in a newborn with large PDA before, during, and after successful therapy.

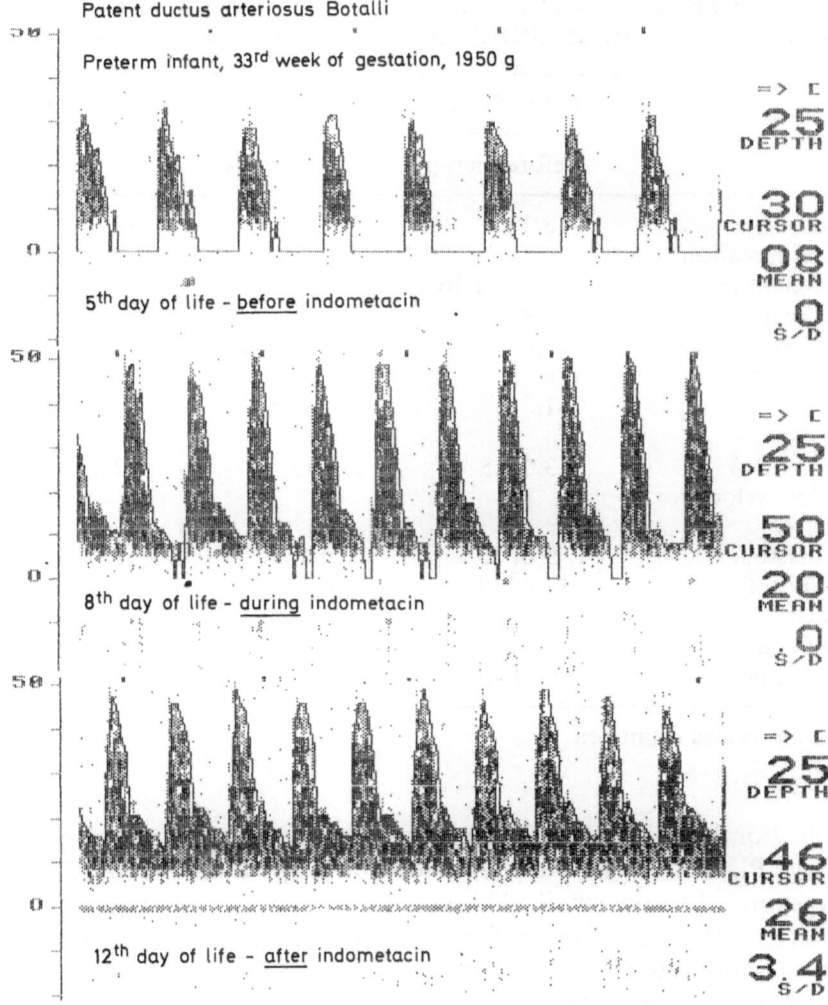

Fig. 22. Doppler signals from the MCA of a premature infant of 33 weeks gestational age with patent ductus arteriosus—before, during, and after therapy

In premature newborns with a large PDA significantly lower flow velocities were found before therapy than after. The reduction of vd was particularly noticeable. The flow velocities in children with a small PDA corresponded to the those in children with a large PDA after successful therapy (Table 10).

Since flow velocities of less than 6 cm/sec could only be estimated (Chapter 3.1.), an exact determination of the vd, SD, and RI was no longer

Table 10. Doppler recordings from the MCA of premature newborns with patent ductus arteriosus Botalli (PDA)—mean values and standard deviations

	Large PDA		Small PDA
	Before therapy	After therapy	
Age (days)	13/12	20/13	17/8
Week of gestation	29/2		32/3
Birth weight (g)	1,163/347		1,566/368
vs (cm/sec)	38/9	52/11	53/14
vm (cm/sec)	13/4	25/8	25/7
vd (cm/sec)[a]	<6	10/5	9/5
RI[a]	1	0.82/0.09	0.83/0.12
SM (=vs/vm)	3.0/0.5	2.1/0.3	2.2/0.3
Peak flow velocity <6 cm/sec			
—pandiastolic	9 of 13	0 of 13	1 of 7
—enddiastolic	13 of 13	3 of 13	2 of 7
Convex-shaped drop of peak flow velocities	9 of 13	0 of 13	1 of 7

[a] Some values estimated

possible. For this reason SM = vs/vm was introduced, which in the presence of a large PDA was clearly higher prior to therapy than afterwards, when it corresponded to the small PDA values. SM was higher than 2.5 in 11/13 children with a large PDA before therapy, in 1/13 children after therapy, and in 1/7 children with a small PDA (Table 10). In contrast, SM >2.5 in only 8/147 healthy preterm and term newborns was found after the third day of life.

The diagnosis of a large PDA was established with TCD, independent of the clinical findings, when at least 3 of the following criteria were fulfilled:
1. enddiastolic peak flow velocity <6 cm/sec
2. pandiastolic peak flow velocity <6 cm/sec
3. SM >2.5
4. convex-shaped drop of peak flow velocities after the systolic peak.

Upon changes in ventilation or vigilance, the flow velocities in these children fluctuated less than they did in comparable children without a large PDA. More than one criterion was hardly ever met in children with a small PDA. The clinical and Doppler sonographic differential diagnoses of large and small PDA correlated well (Table 11).

Table 11. Correlation of clinical and Doppler sonographic differential diagnosis of a large/small PDA ($p < 0.05$ in χ^2 test)

		Doppler sonographic diagnosis	
		large	small
clinical	large	12	1
diagnosis	small	1	6

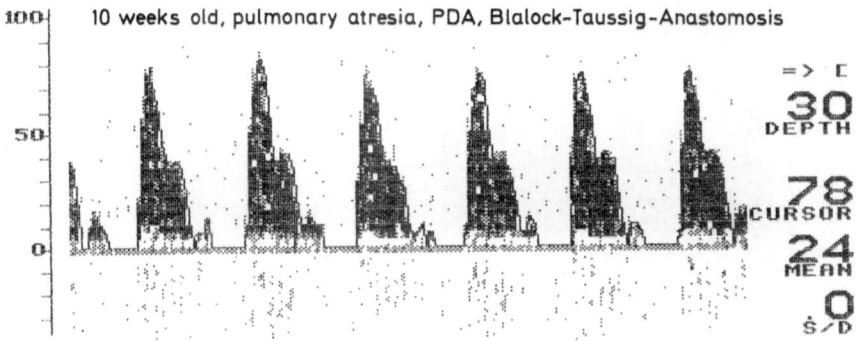

Fig. 23. Doppler signals from the MCA of a 10 week old infant with pulmonary atresia, open ductus arteriosus, and Blalock-Taussig anastomosis

In examinations carried out on consecutive days it was often observed that the incidence of the typical criteria for a large PDA changed from day to day. In the presence of hypercapnia the PDA diagnoses could frequently not be made.

In two out of six infants with a large PDA and additional heart defects, vd below 6 cm/sec was found. In a 10 week old baby with pulmonary atresia, PDA, and Blalock-Taussig anastomosis, Doppler sonography showed all the criteria for a large PDA (Fig. 23).

Three newborns with a PDA as part of a persistent fetal circulation syndrome [347, 362, 404] showed no indication of a PDA upon Doppler examination.

4.4.2. Perinatal brain damage

Perinatal hypoxia

In hours-old newborns with severe perinatal hypoxia, the flow velocities were increased. This particularly applied to the vd, so that the SD and RI were much lower than in comparable healthy newborns. The abnormalities

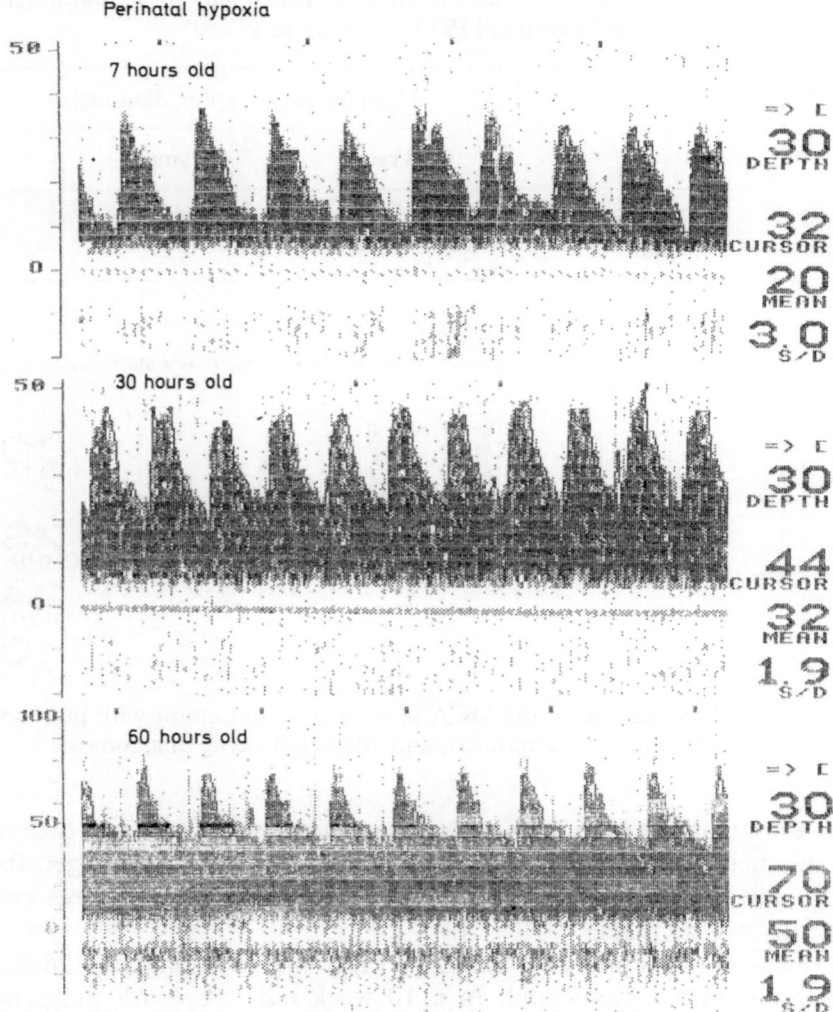

Fig. 24. Doppler signals from the ICA of a newborn after perinatal hypoxia—measurements 7, 30, and 60 hours after birth

persisted despite hyperventilation therapy. On the 2nd to 3rd day of life, the velocities increased to twice the reference values (Fig. 24). In the cases with favorable outcome they returned to normal during the next days.

The fatal cases took different courses, depending on the supportive measures:

— in non-ventilated children the flow velocities were markedly raised in the phase immediately preceeding apnea. After breathing had ceased and in the presence of bradycardia we saw characteristic Doppler signals

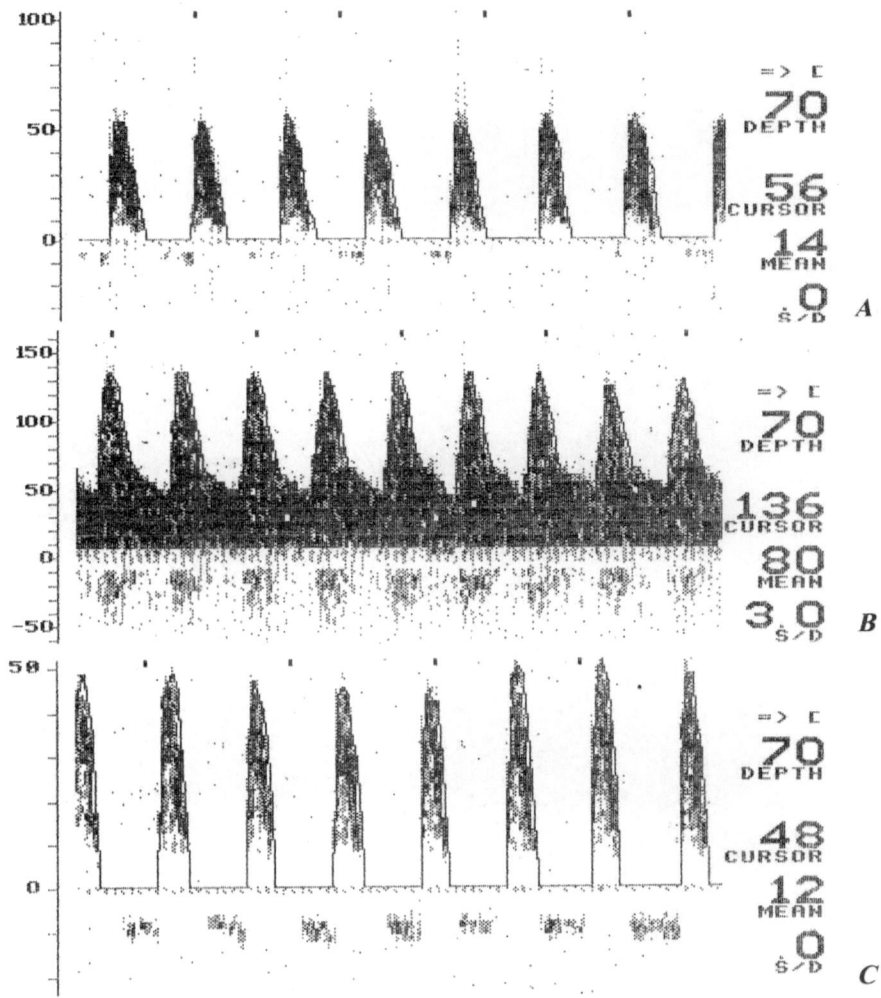

Fig. 25. Doppler signals from the ICA of a newborn after perinatal hypoxia—measurements during arterial hypotonia (**A**), after intratracheal administration of epinephrine (**B**), and during clinical signs of brain death (**C**)

with a steep increase in flow velocities in systole and exponential drop to very low enddiastolic values. The flow velocities decreased with the length of the respiratory standstill (Fig. 15, Chapter 4.2.7.).

— in mechanically ventilated babies the flow velocities underwent pressure passive changes just before death (Fig. 25). An arterial hypotonia resulted in a marked decrease of the velocities. After the intratracheal administration of epinephrine the velocities increased sharply for a short time. Finally, with the cardiac action still regular, a reverberating pattern

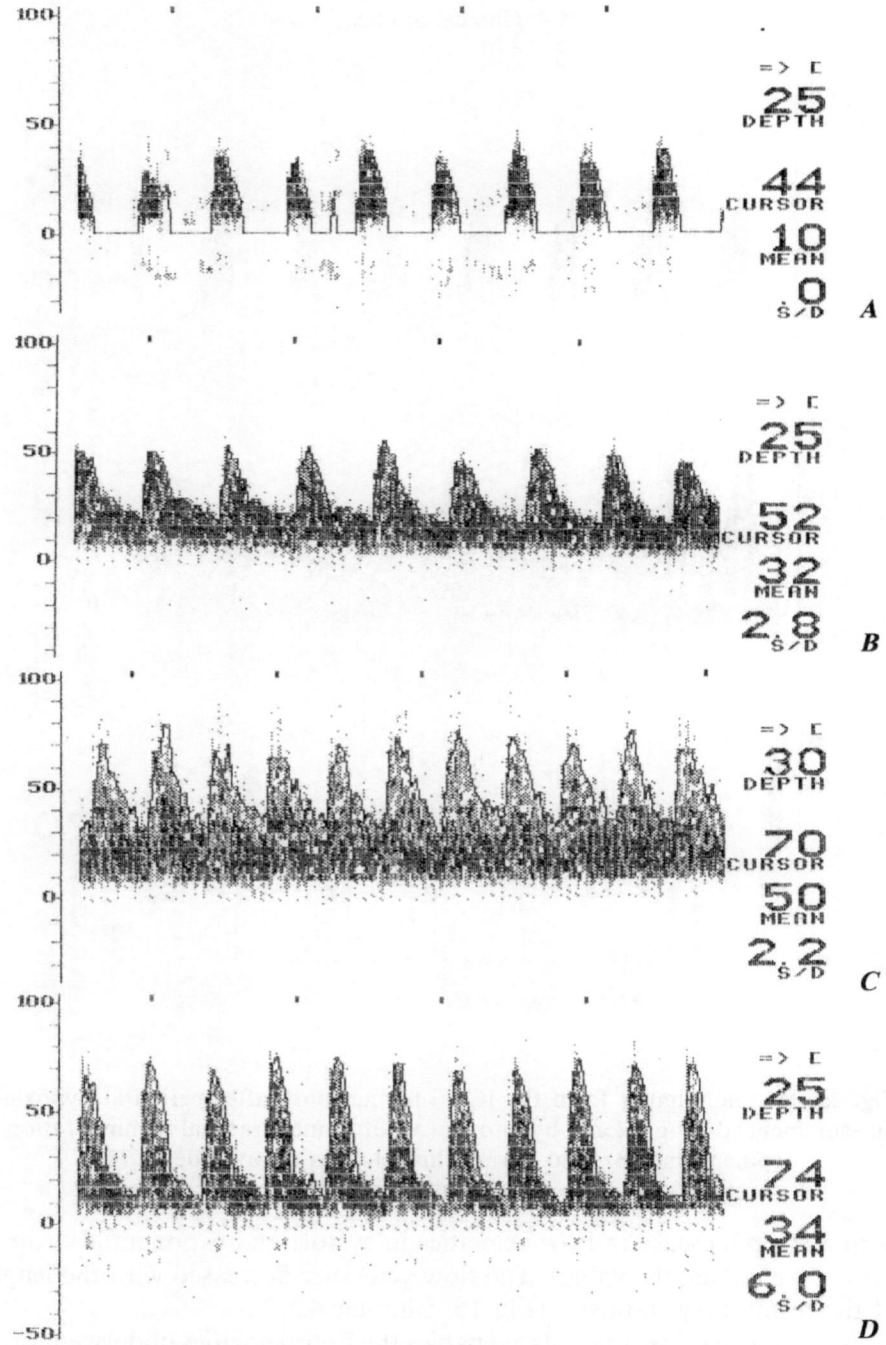

Fig. 26. Right cerebral infarction in a small-for-date baby; blood flow velocities in the left and right MCA and 2nd (**A**), 4th (**B**), 7th (**C**), and 18th (**D**) day of life; p. 52—left MCA, p. 53—right MCA

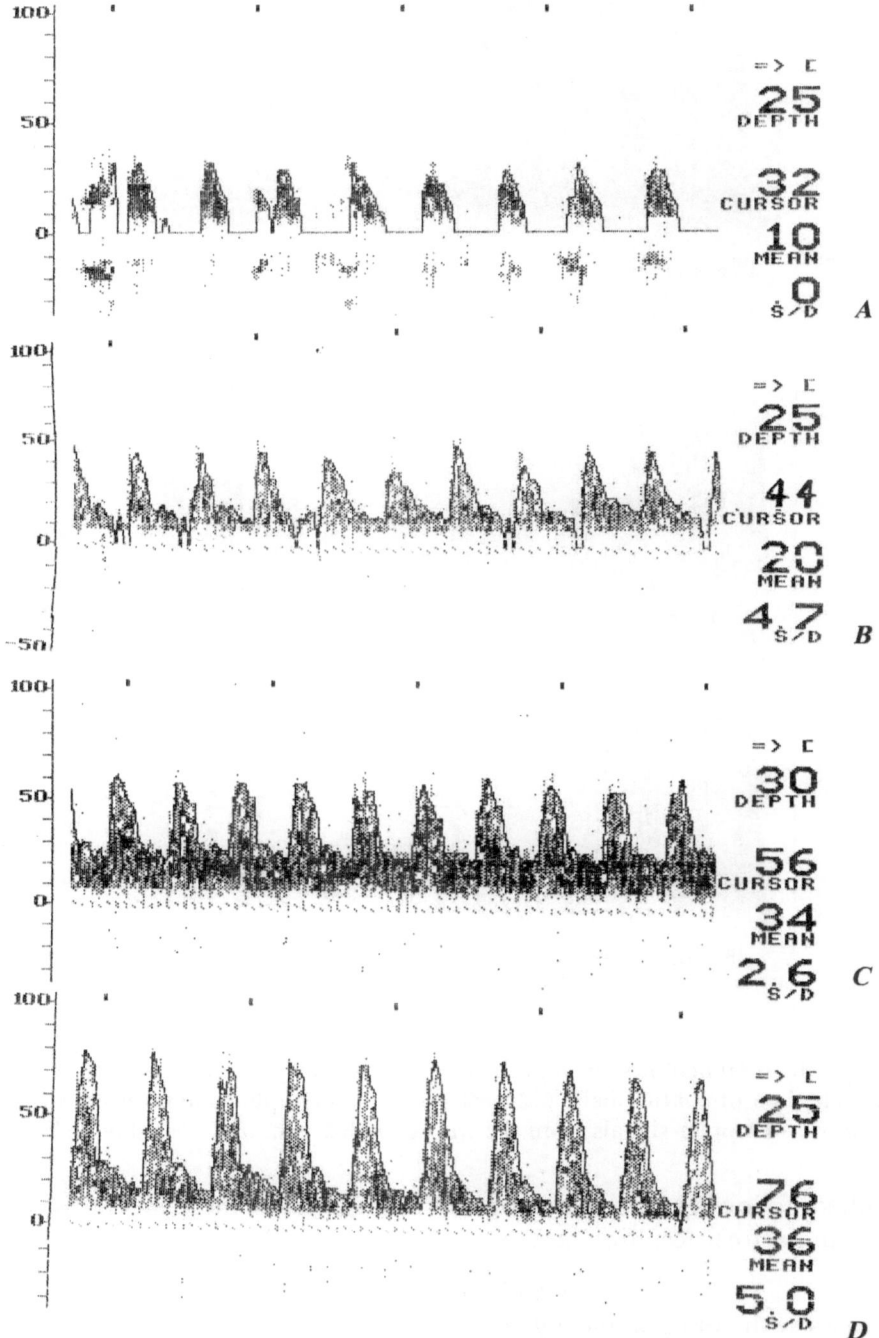

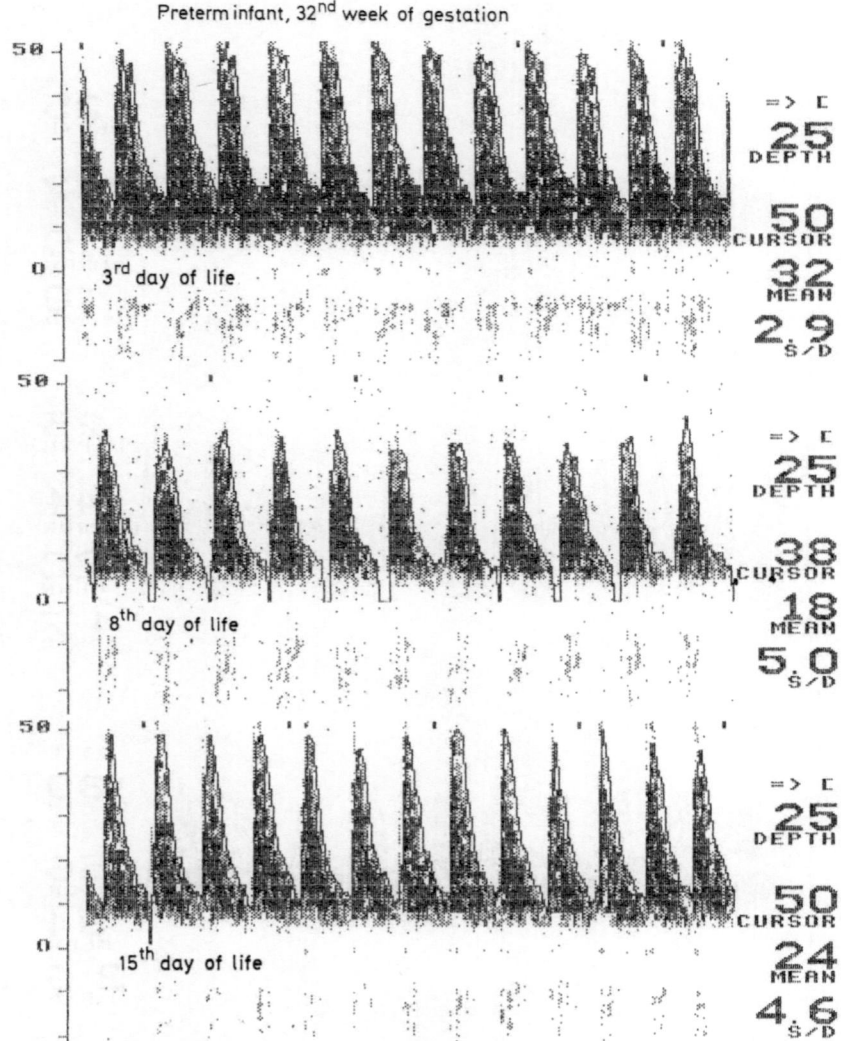

Fig. 27. Bilateral hemorrhage into the cerebral parenchyma with invasion into the
ventricles in a premature baby (32 weeks GA), sonographic diagnosis on 3rd day
   of life—Doppler signals from the MCA on 3rd, 8th, and 15th day of life

resulted: following a sharp orthograde flow in systole there was retrograde
flow in diastole, resulting in a zero net flow.

Cerebral infarction

In cases with sonographically confirmed cerebral infarction we observed
diminished flow velocities on the side of the infarction for a period of more
than 7 days (Fig. 26).

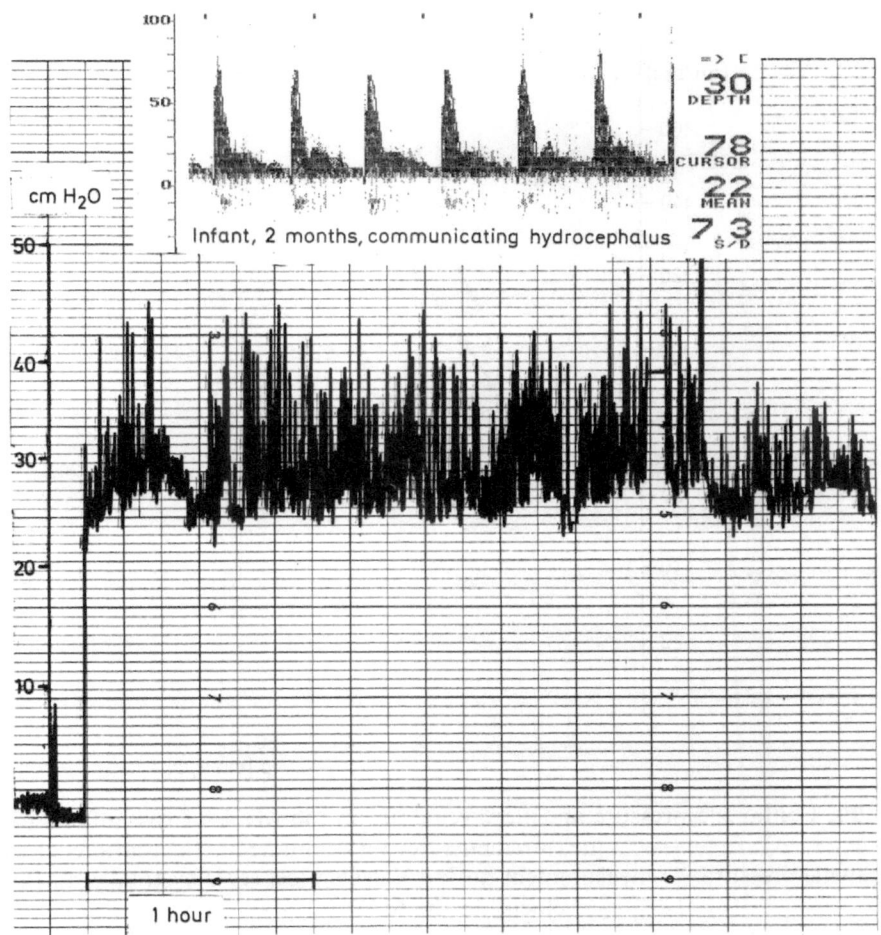

Fig. 28. Simultaneous measurement of the fontanelle pressure and blood flow velocities in the MCA of a 2 month old infant with hydrocephalus communicans

Cerebral hemorrhage

On the day extensive cerebral hemorrhage was diagnosed by B-scan, increased flow velocities were observed. During the following days, all of the flow velocities, particularly the vd, were reduced on both sides. The raised values of SD and RI persisted when a posthemorrhagic hydrocephalus developed (Fig. 27).

4.4.3. Increased intracranial pressure

The relationships between increased intracranial pressure and blood flow velocities were investigated in two studies:

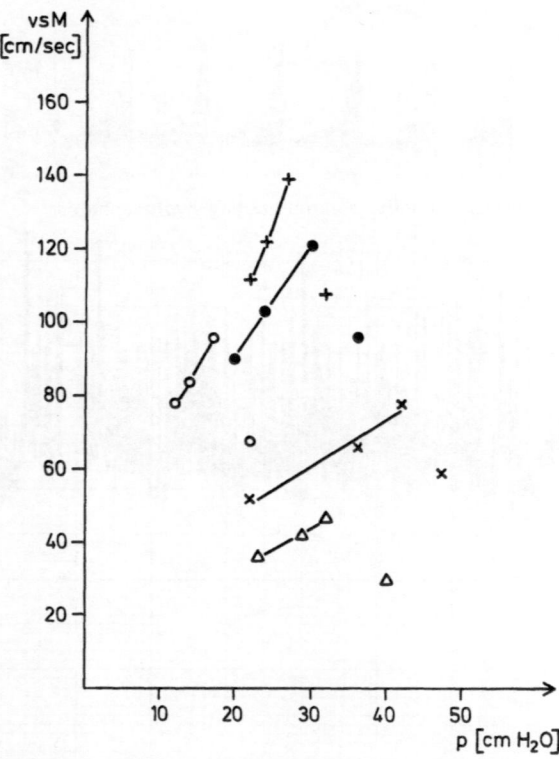

Fig. 29. Fontanelle pressure and flow velocities in the MCA of five infants (○, +, ●, ×, Δ): ○────────○ measurements during spontaneous fluctuations of the fontanelle pressure. ○ measurement during Valsalva test (*vsM* systolic peak flow velocity in the MCA)

Fluctuations of the fontanelle pressure in infants

The fontanelle pressure was measured in five infants with hydrocephalus using applanation tonometry [101, 128, 172, 180, 267, 281, 316, 317, 341, 363, 370, 390, 391] with a LADD M 1000 pressure monitor. The flow velocities in the MCA were determined simultaneously. In all of the children undulating pressure fluctuations around a constant mean value were found. There were no typical pressure waves [127–129, 211, 307], as is shown in Fig. 28. Apart from the clearly raised fontanelle pressure, a low vd and a high SD with normal vs is noticeable.

The flow velocities in all of the children were higher when the fontanelle pressure increased. During a Valsalva test they dropped with increasing fontanelle pressure (Fig. 29).

Table 12. Doppler parameters in the MCA (in cm/sec) and intracranial pressures (p, in cm H_2O before (*1*) and after (*2*) decompression

Age	vs/1	vs/2	vm/1	vm/2	vd/1	vd/2	p/1	p/2	RI/1	RI/2
5 weeks	64	84	30	52	14	28	20	8[a]	0.78	0.66
3 months	82	88	42	50	16	30	22	8[a]	0.80	0.66
16 months	106	106	76	72	50	50	23	8[a]	0.53	0.53
3 weeks	58	64	24	36	4	22	24	8[a]	0.93	0.66
3 weeks	74	88	38	42	12	20	13	8[a]	0.83	0.77
11 months	80	88	44	52	26	32	24	8[a]	0.68	0.64
24 months	118	130	80	100	65	65	45	7[b]	0.45	0.50
4 weeks	100	92	58	56	24	30	32	16[b]	0.76	0.67
5 weeks	92	120	64	72	42	48	22	8[a]	0.54	0.60
6 months	100	104	64	68	40	44	13	8[a]	0.60	0.58
14 years	100	106	70	68	50	54	14	10[a]	0.50	0.49
4 months	118	110	72	68	42	46	30	8[a]	0.64	0.58
6 weeks	66	62	34	32	14	14	22	10[b]	0.79	0.77
3 months	96	100	54	58	22	30	18	8[a]	0.77	0.70
2 months	50	76	16	36	8	16	41	26[b]	0.84	0.79
Mean values	87	95	50	57	29	35	24	10	0.70	0.64
Sign test	—$p<0.05$—		—n.s.—		—$p<0.05$—		—$p<0.05$—		—$p<0.05$—	

[a] Estimated values after shunting
[b] Measured values after ventricular puncture

Change in flow velocities caused by the normalization of previously
increased intracranial pressure

Fifteen children with increased intracranial pressure as a result of hydrocephalus internus were examined before and after decompression (ventricular puncture or shunting). The vs, vm, vd, and RI in the MCA were
determined in each child (Table 12).

There was no correlation of the flow velocities with the level of intracranial pressure. In some of the children the flow velocities increased markedly after decompression, in others they did not. The vs and vd in the
entire group increased significantly after decompression. This tendency,
however, was not so pronounced that the Doppler findings alone would
have permitted the diagnosis of an increased intracranial pressure.

Case reports

Cerebral edema: A 2 year old child had a posttraumatic cerebral edema
on the left side, which was documented in CT. On the day after the trauma

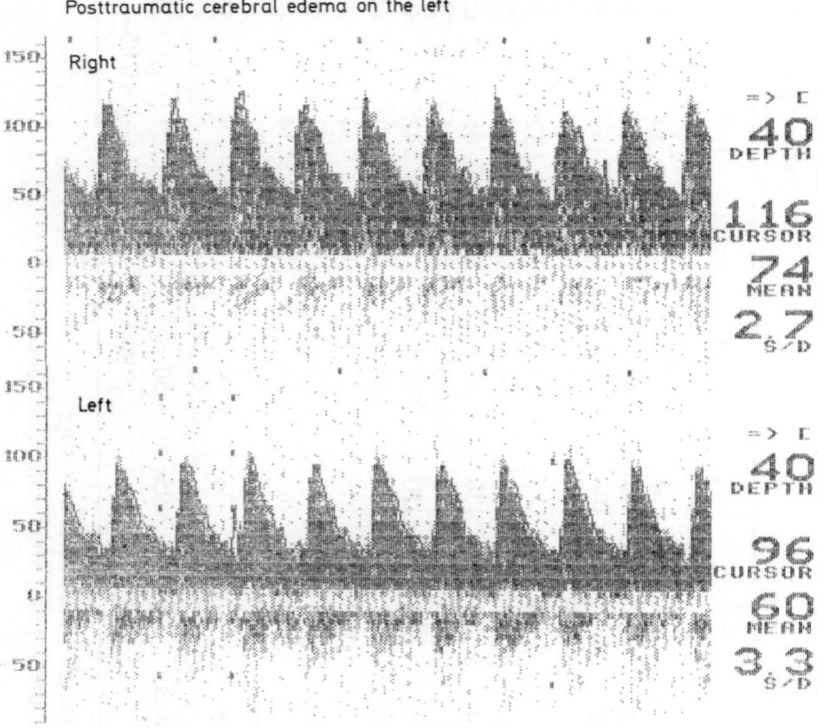

Fig. 30. Posttraumatic left cerebral edema in a 2 year old child—Doppler signals
from the MCA on both sides on 1st day after the trauma

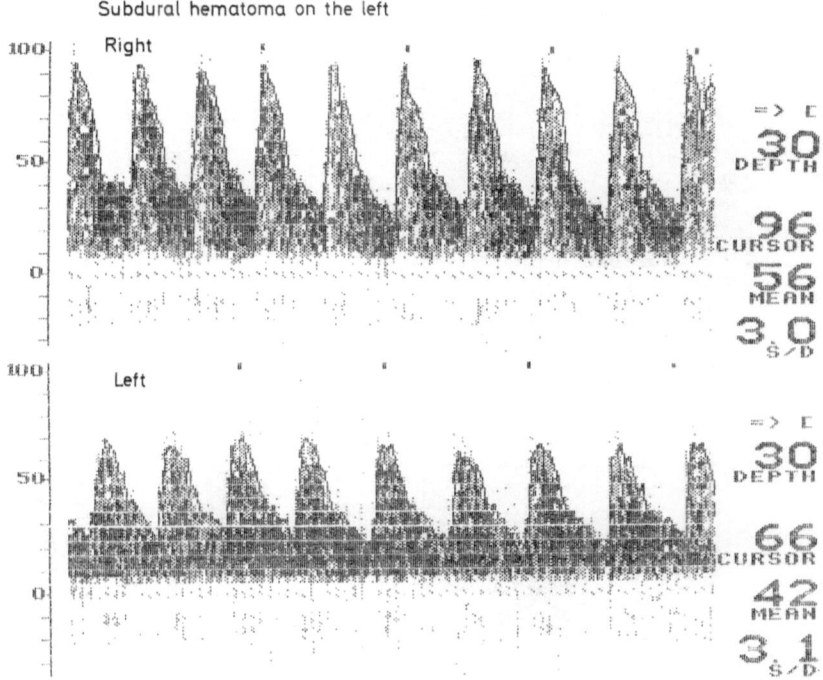

Fig. 31. Subdural hematoma on the left in a 5 month old infant—Doppler signals from the MCA on both sides

lower flow velocities were found on the affected side than on the healthy side (Fig. 30). Within a few days the findings were normal again.

Subdural hematoma: A 5 month old baby with a subdural hematoma on the left showed reduced flow velocities on this side (Fig. 31).

Near drowning: A 12 year old boy who nearly drowned presented with pathological pressure waves in the epidural long term pressure recording. High vd with corresponding low SD were observed (Fig. 32). The flow velocities were higher in the presence of higher pressure values (Fig. 33).

Seizure: During a generalized tonic seizure in a comatose 5 month old infant following a near miss sudden infant death, an increase in fontanelle pressure from 15 to 35 cm H_2O and markedly increased flow velocities were found (Fig. 34).

Plexus papilloma: A 5 month old infant with a plexus papilloma on the left side with a considerable increase in intracranial pressure had significantly higher flow velocities on the side of the tumor (Fig. 35).

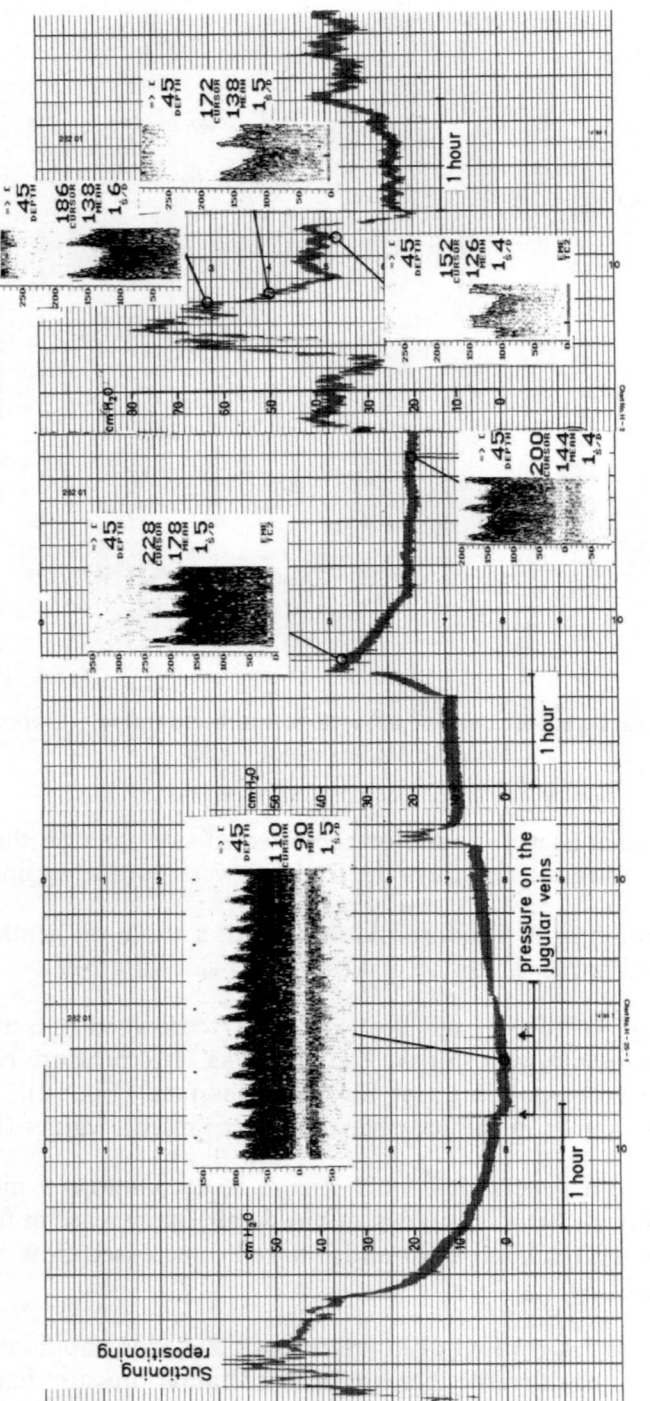

Fig. 32. Epidural long term pressure recording and repeated measurements of the flow velocities in the left MCA in a 12 year old boy after a near drowning

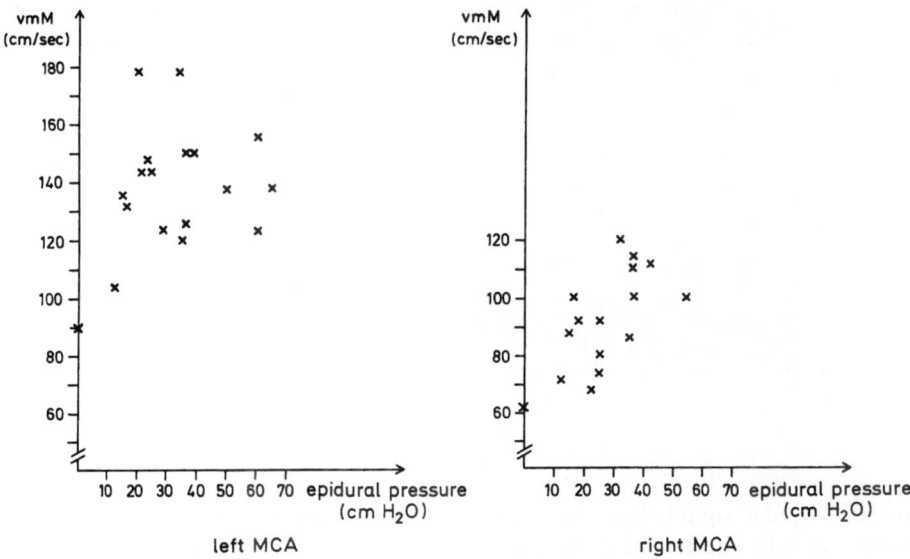

Fig. 33. Relationship between epidural pressure and flow velocities in a 12 year old boy after a near drowning (difference between sides as a result of iatrogenic AV fistula on the right, *vmM* mean peak flow velocity in the MCA)

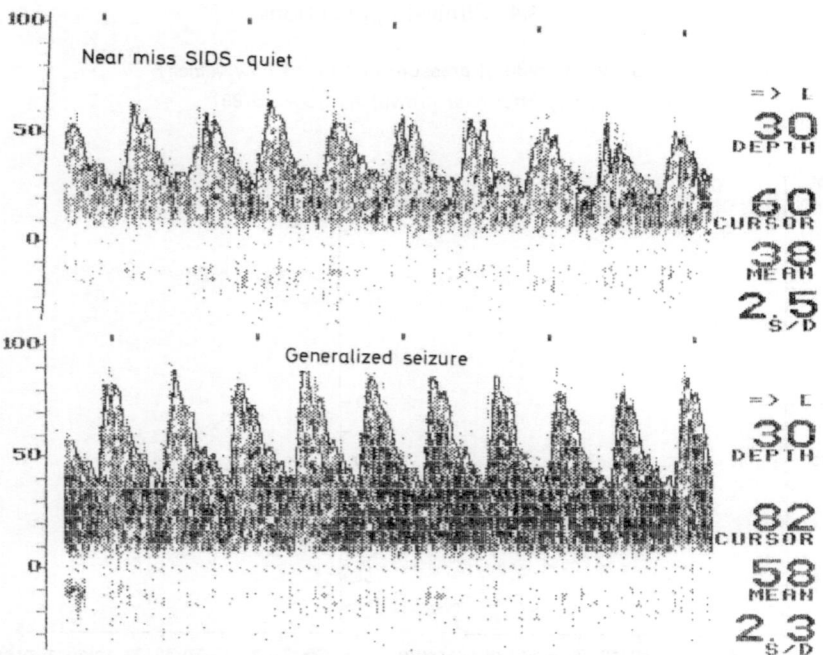

Fig. 34. Doppler signals from the right MCA of a 5 month old comatose infant after a near miss sudden infant death. Measurements performed in quiet state and during a generalized seizure

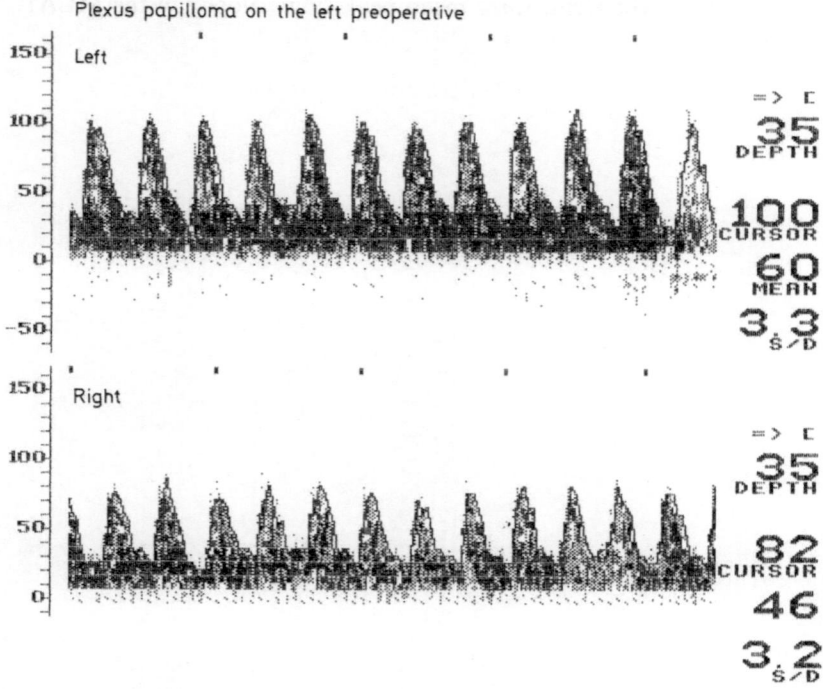

Fig. 35. Plexus papilloma in a 5 month old infant—Doppler signals from the MCA on both sides

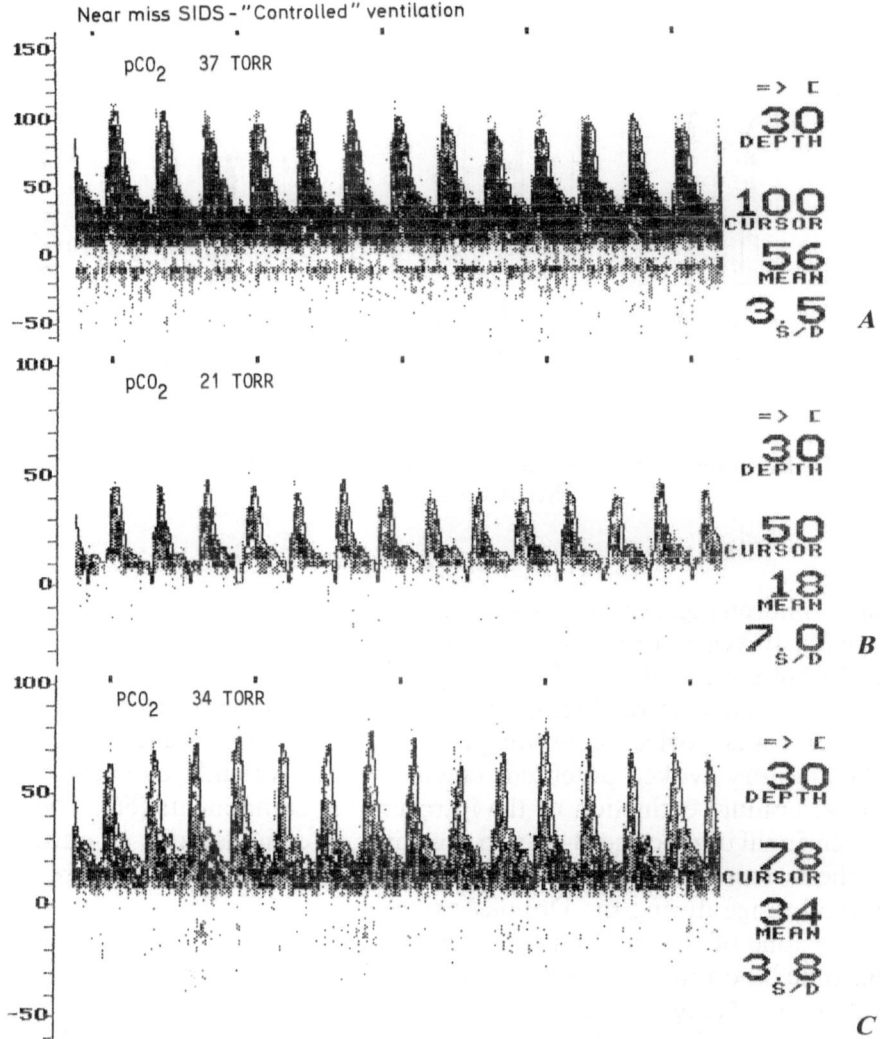

Fig. 36. Hyperventilation therapy of a 5 month old infant after near miss sudden infant death—Doppler signals from the right MCA. The measurement **B** induced a change in the ventilation therapy

Hyperventilation therapy of cerebral edema: Transcranial Doppler sonography was used to monitor hyperventilation therapy of cerebral edema. In this way, insufficient blood supply to the brain as a result of excessive hyperventilation could be detected very quickly (Fig. 36). This was basis for modifying the ventilation therapy. In the presence of pathologically raised flow velocities (Fig. 24), hyperventilation was forced.

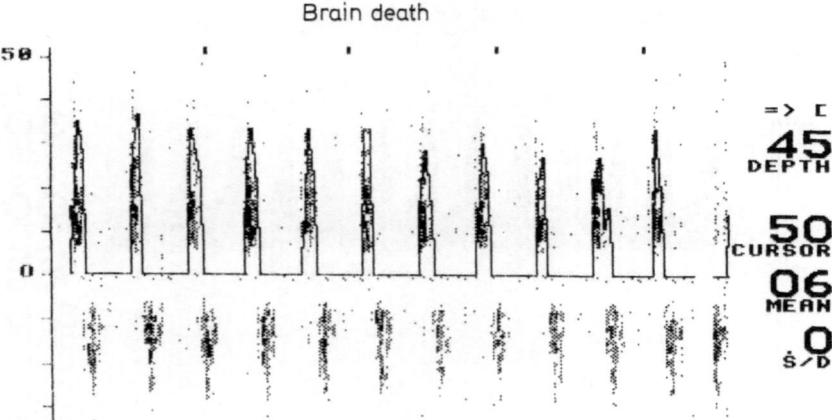

Fig. 37. Brain death in a 10 year old child—Doppler signals from the proximal MCA on the right side

4.4.4. Brain death

Nine children (aged 2 days, 4 days, 10 months, 16 months, 2 years, 3 years, 7 years, 10 years, 12 years) with the clinical signs of brain death [66, 104, 265] were examined.

The electroencephalogram performed on seven cases in the intensive care unit was isoelectric, although there was a high incidence of artifacts. The auditory evoked potentials determined in five children showed a bilateral gradual extinction of the intracerebral components [66]. The two-dimensional ultrasound B-scan in three infants demonstrated no pulsations of the intracranial arteries. The arterial blood pressure values were in the normal range during the Doppler examination.

In none of the children was there an intracranial orthograde systolic-diastolic blood flow. A reverberating pattern in both ICA's and in the BA was recorded (Fig. 25, Chapter 4.4.2.). When the ventilation was continued, only brief, weak Doppler signals alternating in direction between systole and diastole could be detected (Fig. 37).

4.4.5. Cerebral malformations

Hemimegalencephaly

Greatly reduced flow velocities on the right side were found in a 10 week old infant with a right sided hemimegalencephaly. Eight weeks later the child presented with diffusely mixed seizure activity in the electroencephalogram and increasing psycho-mental deterioration. The flow velocities were reduced on both sides, being only half as high on the right as on the left (Fig. 38).

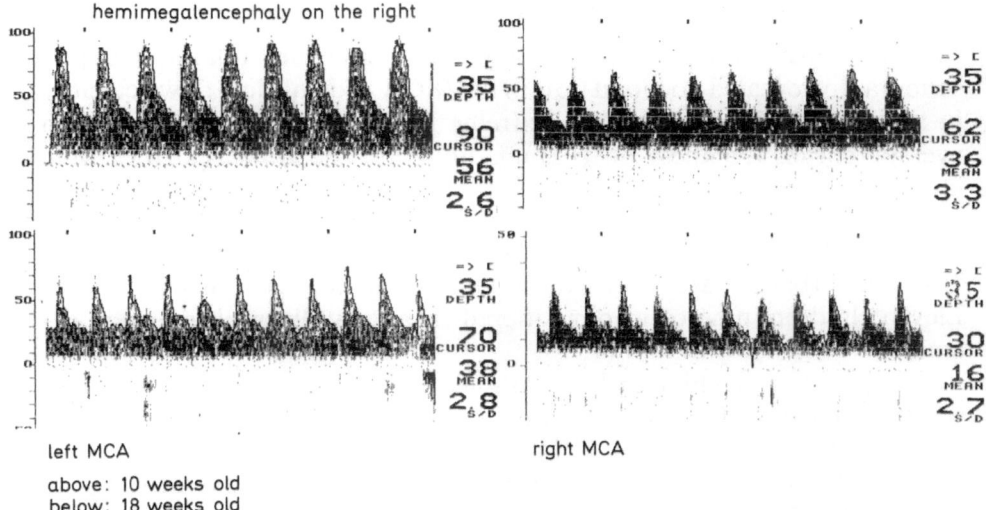

Fig. 38. Hemimegalencephaly on the right—Doppler signals from the left and right MCA at the ages of 10 and 18 weeks

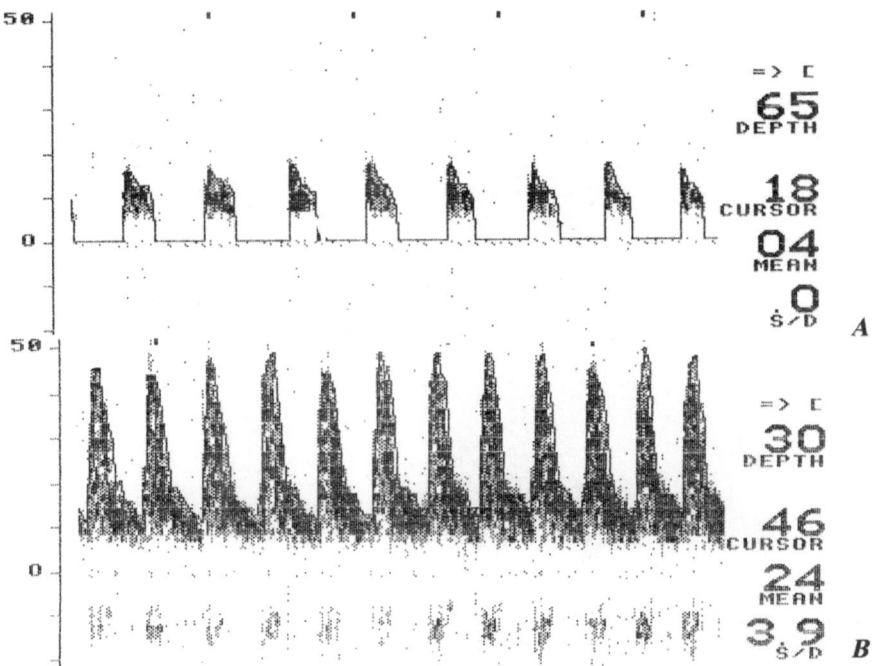

Fig. 39. Doppler signals in **A** hydranencephaly (right ICA), **B** extreme hydrocephalus internus (right MCA)

Hydranencephaly

In a newborn with hydranencephaly and cerebellum-aplasia, Doppler sonography revealed no blood flow in the MCA and only low flow velocities in the ICA. In a newborn with extreme hydrocephalus internus and an average cortex thickness of 1–2 mm, however, the flow velocities were initially normal for the age (Fig. 39).

4.4.6. Cerebral circulation impairments of vascular origin

These disturbances are quite rare in children in comparison with adults. Eleven children, nine of whom suffered from acute hemiplegia, were examined.

Nine of the 11 children had pathological findings. Angiography was carried out in four children. Seven out of nine children with acute hemiplegia had pathological findings in TCD.

In four out of nine children with hemiplegia slower flow velocities were found in the MCA on the affected side than those on the other side. CT showed infarctions in the area supplied by these vessels. The velocities in the ICA were the same on both sides.

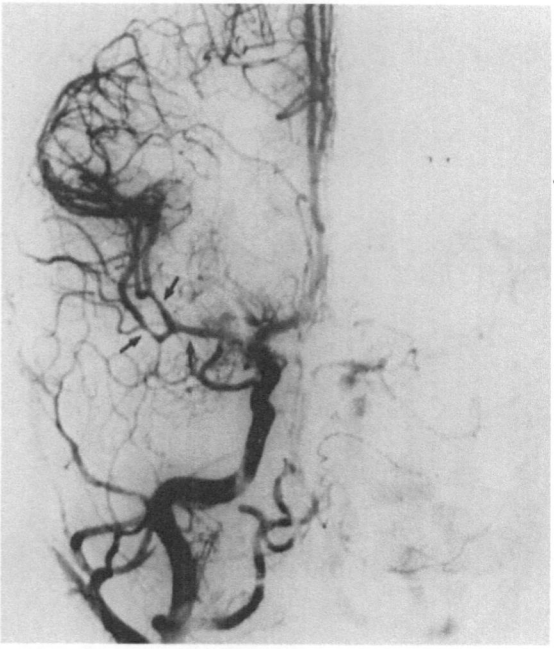

A

Fig. 40. Vasospasm in a 14 year old boy after subdural hematoma. **A** Brachialis angiography on the right: narrowing of the entire right median trunk. **B** Doppler signals from the right MCA on 9th postoperative day (**a**) and in 11th postoperative week (**b**)

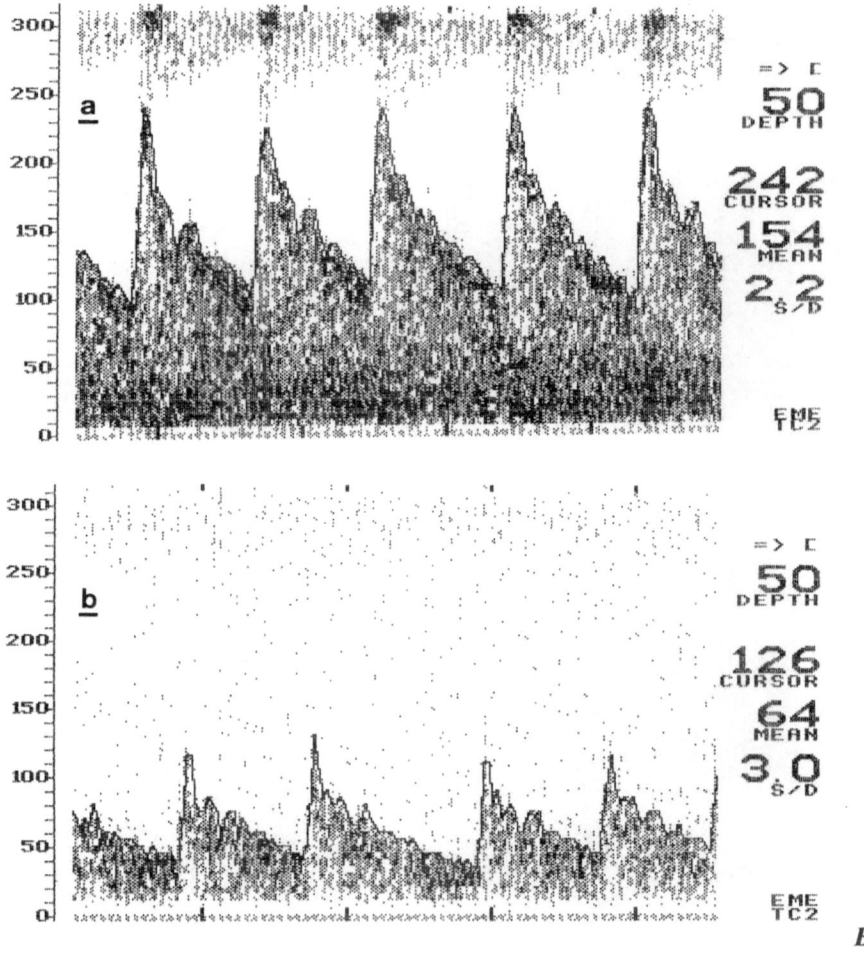

Fig. 40 B

In one child a fast retrograde blood flow could also be detected at an atypical site. Angiography showed an incomplete occlusion of the left MCA and a small collateral artery with a retrograde course. One child had a left-sided hemiplegia as a result of a hemorrhage from a 2 cm large angioma in the right internal capsule that could not be visualized well in angiography. Here somewhat higher velocities in the right MCA were found as compared with those on the left.

Two of the nine children with hemiplegia showed increased flow velocities in the corresponding arteries. Angiography performed in one of these children revealed multiple vessel stenoses.

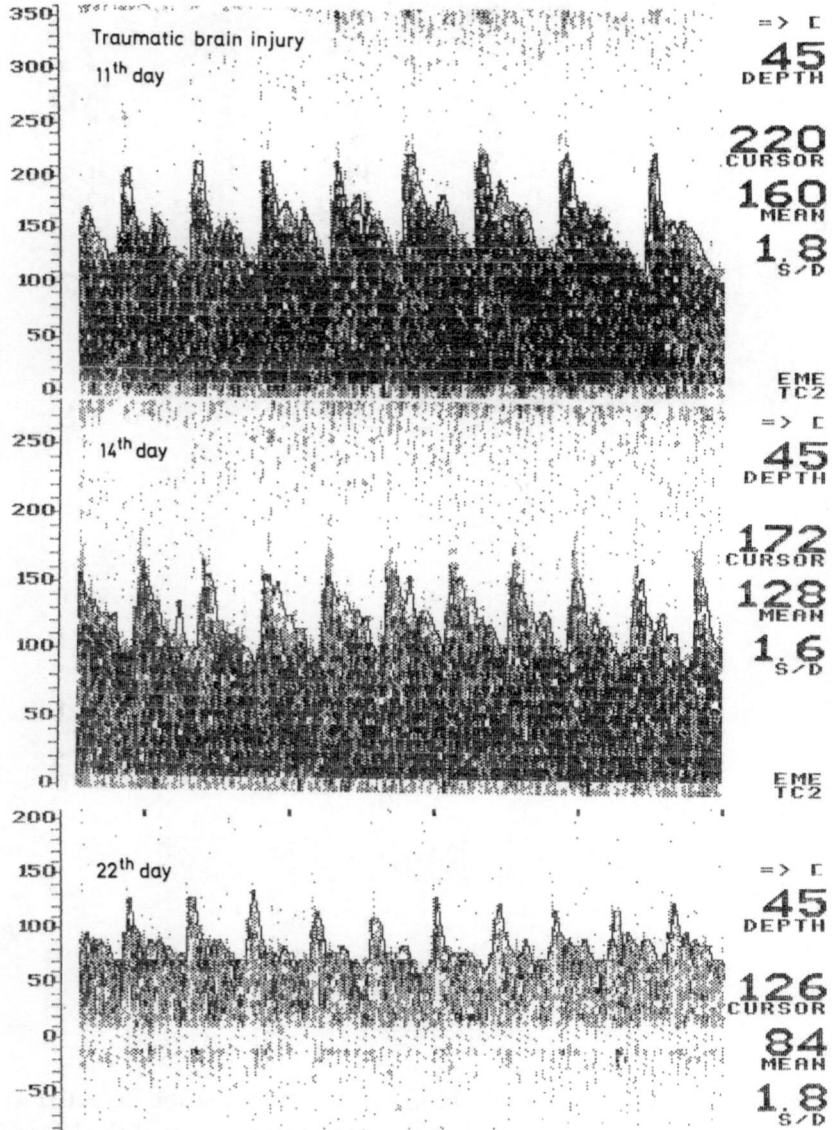

Fig. 41. Vasospasm in a 6 year old boy after traumatic brain injury—Doppler signals from the right MCA on days 11, 14, and 22 after the trauma

Two children without hemiplegia showed temporarily raised flow velocities as a sign of vasospasm, which in one case was confirmed by angiography.

Case reports

Vasospasm: 1. In a 14 year old boy with normal clinical findings angiography revealed multiple narrowings in the main trunk of the right MCA 10 days after a right-sided subdural hematoma had been surgically removed (Fig. 40A). On the 9th postoperative day, the flow velocities in the right MCA were increased to 2.5 times the norm for age. A control examination performed 10 weeks later showed normal values (Fig. 40B). In the left MCA the velocities were always normal for age.

2. A 6 year old boy suffered multiple intracranial contusion hemorrhages and generalized cerebral edema after a traffic accident. Markedly increased flow velocities and low SD ratios were observed in both MCA's and ICA's 11 and 14 days after the accident. The vd was about twice the norm for age. The velocities in the BA were normal. On the 22nd day after the accident, it was found that the velocities in the MCA and ICA were normal again (Fig. 41). The child suffered mesencephalic syndrome for months.

Moyamoya disease: In a 9 year old girl with acute right-sided hemiparesis and a severe left-sided focus in the EEG after hyperventilation, extremely high flow velocities were found in the area where the of MCA branches from the ICA (Fig. 42A). "Musical murmurs" [3] could be heard. Velocities were increased on the right side in the MCA, ICA, and ACA (Fig. 42B). An extreme stenosis at the junction of the left MCA, as well as stenoses of the right MCA, the intracranial carotid bifurcation, and the right ACA

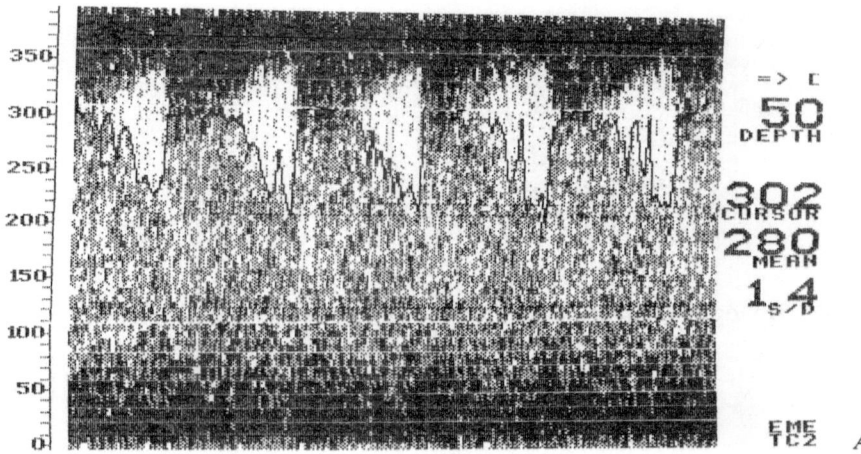

Fig. 42. Moyamoya disease in a 9 year old girl. **A** Doppler signals from the left MCA. (Figs. 42 B–D see pp. 70–72)

4. Results

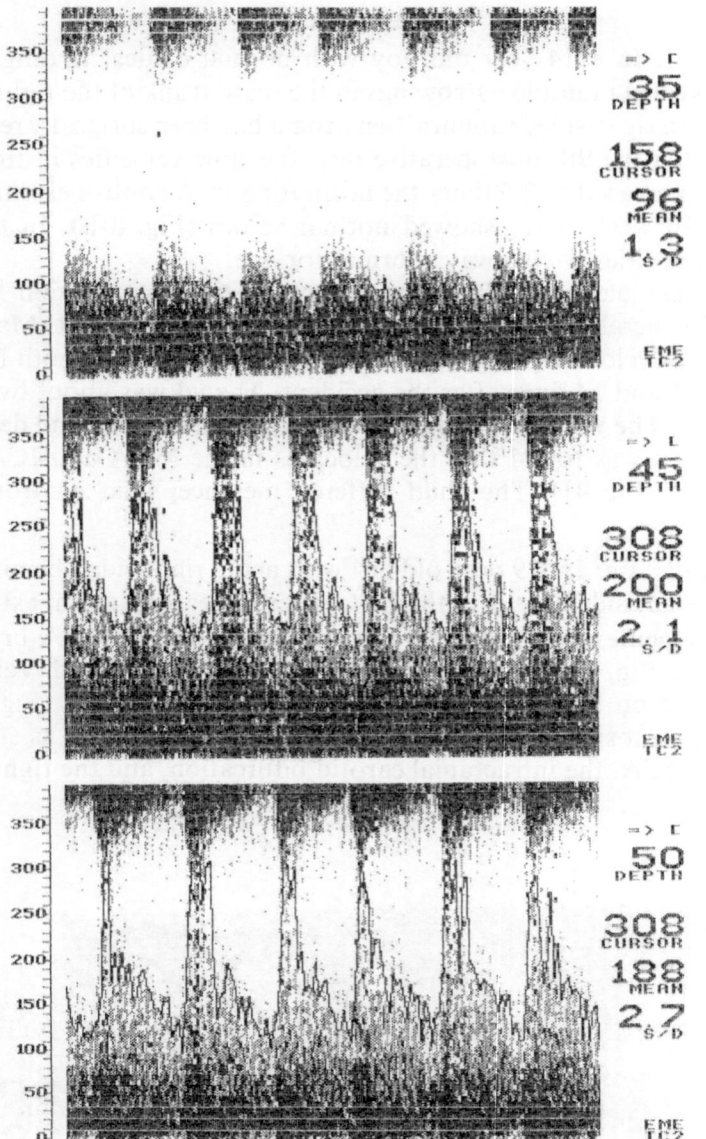

Fig. 42. B Doppler signals from the right MCA (35, 45, 50), ICA (55), ACA (60, 65), and PCA (60)

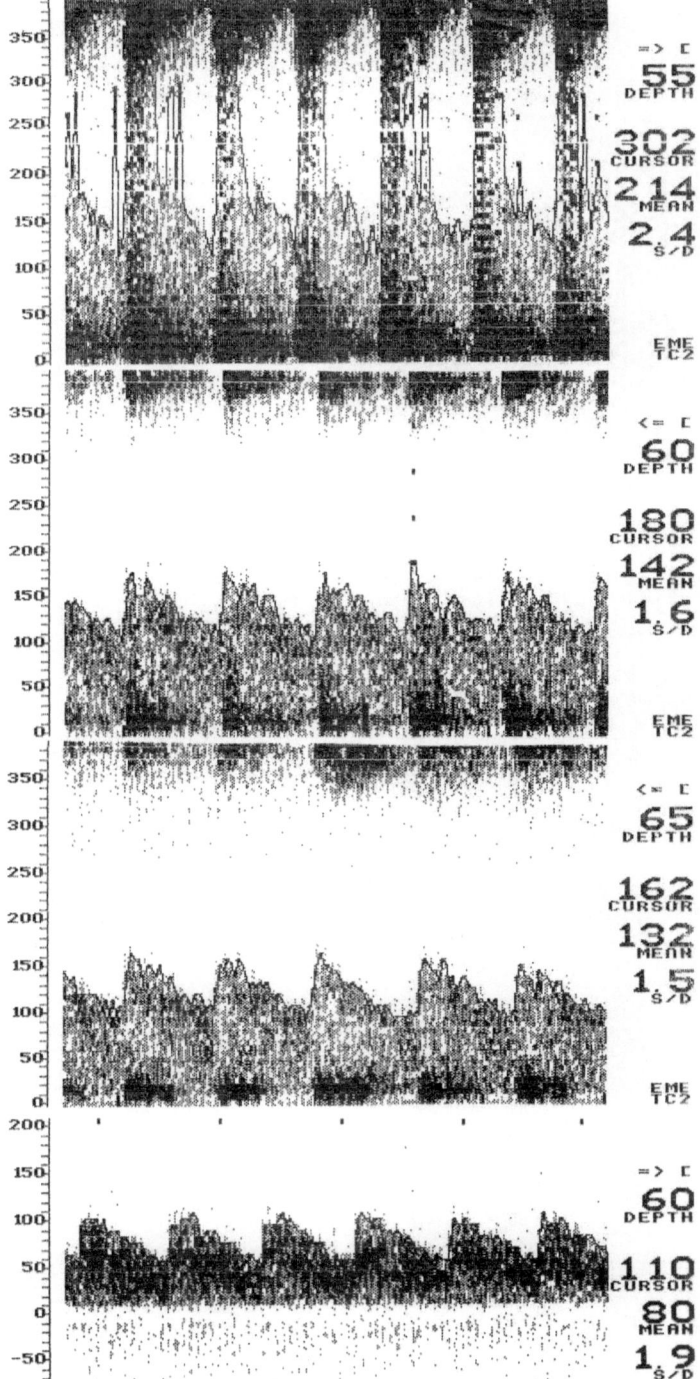

Fig. 42 B

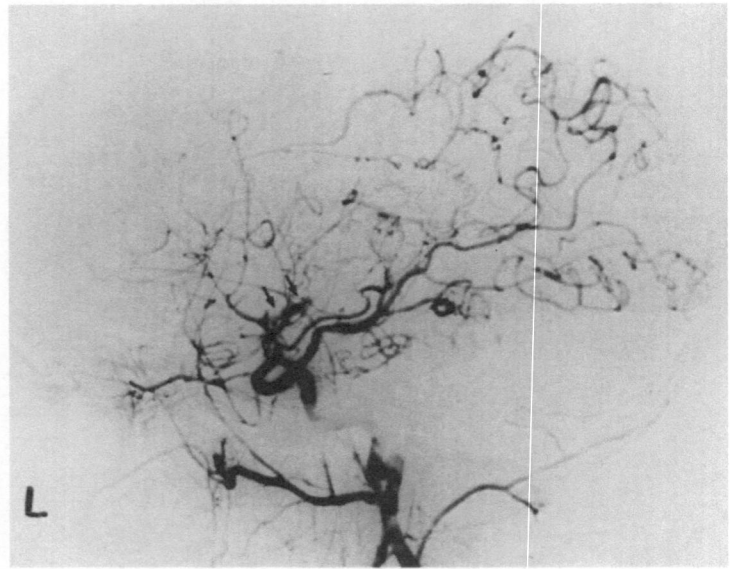

C

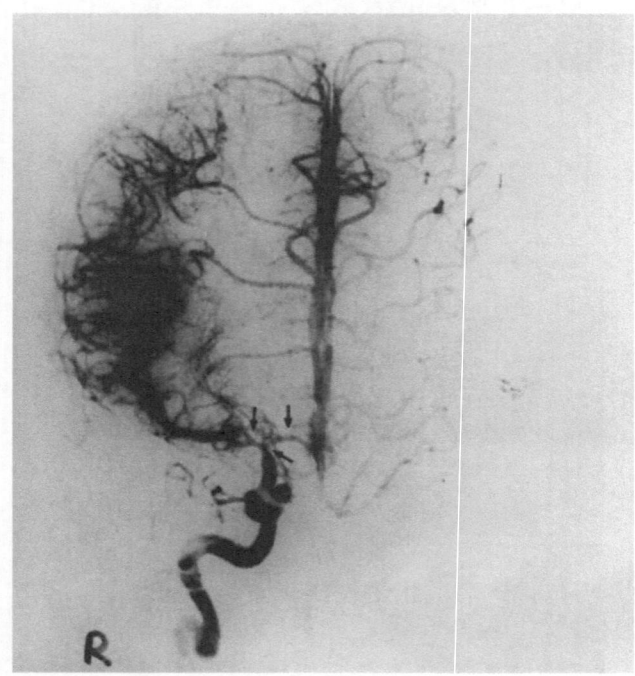

D

Fig. 42. C Carotid angiography left side—high grade stenosis of the left MCA
(appears here as occlusion). **D** Carotid angiography right side—stenosis of MCA,
ICA, and ACA

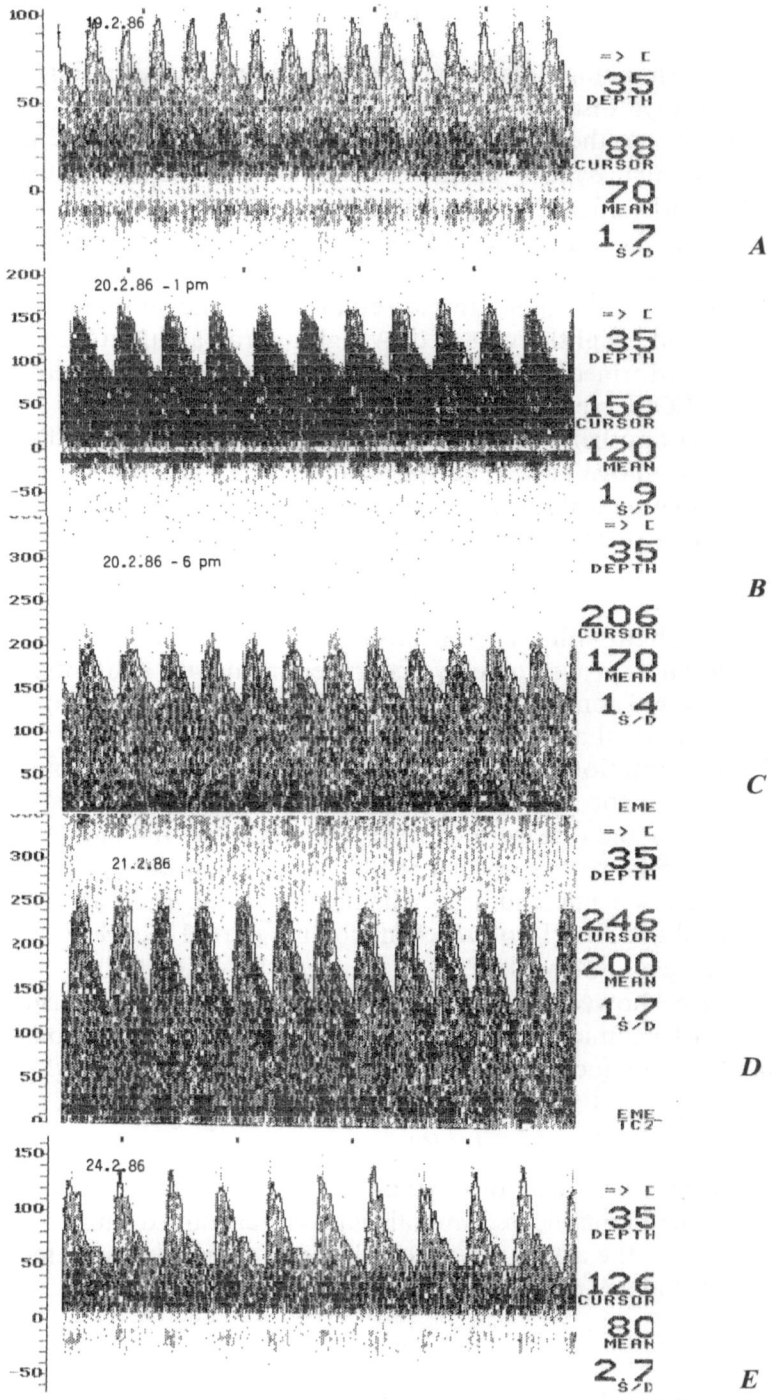

Fig. 43. Pneumococcal meningitis in a 6 month old infant. Doppler signals from the right MCA on days 3, 4, 5, and 8 of the disease

could be seen in angiography (Fig. 42C and D). These findings are typical for Moyamoya disease.

Using TCD, the patency and the hemodynamic effects of an extracranial-intracranial bypass on the left side could be demonstrated for more than 6 months after the bypass had been performed. Later, reduced flow velocities in the right ICA, MCA, and ACA were found with damped waveforms of the Doppler signals. At this point the EEG showed a severe focus on the right. Angiography revealed that the vessel stenoses on the right had only slightly increased. A right extracranial-intracranial bypass was also performed. TCD documented good patency. The focus on the right in EEG decreased postoperatively.

In 1987 another 9 year old girl with occlusions of both distal MCA's due to Moyamoya syndrome was examined. TCD showed on both sides reduced flow velocities in the proximal MCA and no flow in the distal MCA. The flow velocities in the PCA were increased.

4.4.7. Bacterial meningitis

Twelve children with bacterial meningitis were examined. The condition of 9 children improved quickly under treatment with antibiotics and all of them recovered completely. In these children slightly raised vd were found in all of the basal arteries in the first days of the disease.

Despite antibiotic treatment, three infants remained in coma (Glasgow Coma Scale 6) for several days. The flow velocities increased on the 3rd to 5th day of the disease to 3 times the norm for the age (Fig. 43). The fontanelle pressure was normal at that time. A few days later, B scan and CT revealed ischemic and hemorrhagic infarctions in the area supplied by the affected arteries. Two babies made only partial recovery, one neonate died at the age of 4 weeks.

Autopsy showed subacute leptomeningitis in the basal cisterns and a multicystic leucomalacia of both hemispheres with diffuse hypoxic-ischemic damage of gangliocytes.

4.4.8. Vasomotor headache and migraine

Ten children with vasomotor headache and 12 children with migraine (5 of which were complicated migraines) were examined during and within 48 hours after the attack. In no child were abnormal flow velocities measured or velocities that were different on each side.

4.4.9. Other diseases

In two young girls with anorexia nervosa the flow velocities in all of the basal cerebral arteries examined were 20% and 30%, respectively, below the lowest reference values for age (Fig. 16, Chapter 4.2.7.).

In two out of four children with intracranial flow murmurs that were diagnosed by chance, the flow velocities in all of the examined vessels were 20% higher than the highest reference values for age.

Four children with sepsis and two with hemolytic-uremic syndrome had high vd with low SD in the arteries examined.

In the remaining patients the clinical diagnoses varied considerably, the Doppler sonographic findings being normal or uncharacteristic.

4.4.10. Therapy control

The significance of transcranial Doppler sonography for therapy control is discussed under the respective diseases (Chapters 4.2.9., 4.4.1, 4.4.2, 4.4.3, 4.4.6).

5. Discussion

5.1. Biological significance of the Doppler parameters

5.1.1. Problems in comparing methods

The results of Doppler sonography have been compared with those of other methods to elucidate the biological significance of the Doppler parameters. There proved to be basic problems associated with such an undertaking. There is no direct way to control the precision of a measurement. Neither the "true" cerebral blood flow nor the "true" blood flow velocities can be determined with any method. The examinations are usually not performed simultaneously, since each of the techniques requires a different amount of time. Changes in the "true" parameters can occur between measurements, which can seriously affect the results.

The methods measure different aspects of the biological parameters. With Doppler sonography, the blood flow velocities in the large cerebral arteries can be recorded continuously. In the area supplied by the ACA, for example, the white matter is overrepresented, and in that of the ICA the distribution of white and gray matter is about the same as in the entire brain. With clearance and autoradiographic methods, however, the blood flow of certain areas of the brain is integrated during a period. This only partly corresponds to the area supplied by the large arteries and in this the gray matter is clearly overrepresented [147].

Each of the different methods for examining the cerebral hemodynamics has both advantages and limitations. There is no "gold standard" [30].

5.1.2. Resistance index and cerebrovascular resistance

Pourcelot [311] introduced the "index de résistance" for Doppler sonographic investigations in adults with peripheral vascular disease as a measure of the peripheral vascular resistance. This index increases from the central to the peripheral arteries of the body [187]. The intracranial arteries have lower vascular resistance than the rest of the arteries in the body [89, 265]. This resistance index—referred to as the "pulsatility index" by Bada [27, 30]—was used to describe the intracerebral hemodynamics of ill premature infants. A high RI was regarded as an expression of vasoconstriction and thus of high cerebrovascular resistance (CVR), while a low RI was a sign of vasodilatation with low CVR. Several authors felt that the RI semiquantitatively measured the cerebral blood flow (CBF) [242, 301–305, 368, 399].

Animal experiments demonstrated that contrary to Bada's assumption, not only the enddiastolic, but also the systolic flow velocities changed with CVR. The CVR then was defined as the ratio between mean arterial blood pressure and mean arterial blood flow velocity [33, 34].

In a comparison between Doppler sonography and the Xenon-133 inhalation method, the CBF correlated far better with the various blood flow velocities than did the RI [147], although there was an inverse relationship between RI and CBF. Theoretical considerations have suggested that the RI does not seem ideally suited for distinguishing among the various biological states. RI can be identical, even when the vs and vd have completely different values [8, 30, 34, 39, 40, 399]. The same applies for the SD, Gosling's pulsatility index (PI), and the SM ratio (Chapter 2.4.4. and 4.4.1.). The RI is influenced by a host of factors (e.g. blood pressure amplitude, shape of the blood pressure curve, heart rate, blood viscosity, CVR, cerebral blood volume [241], autoregulation and CBF). It is not accepted today as adequately describing the CBF. Where it is useful, however, is in evaluating the CVR [30, 34, 39–42, 399]. The same applies to the SD ratio because of its direct relation to RI (equation 16, Chapter 2.4.4.). Many studies performed in the early 1980's which determined only the RI now have to be revised.

5.1.3. Blood flow velocities and cerebral blood flow

Relationships between the different Doppler parameters and the cerebral blood flow have come to be recognized in recent years.

Experimental investigations

Rosenkrantz [355] and Lundell [254] recorded the flow in a plastic catheter and found a very close relationship ($r = 0.99$) between the area under the Doppler curve (AUC) and the flow volume. Hansen [157] compared the continuous-wave Doppler and the microspheric technique in an animal experiment and discovered a close correlation of AUC ($r = 0.86$), vs, vm, vd ($r = 0.72$) with CBF. In the normal range there was a linear relationship, whereas in the pathological range a non-linear relationship between the flow velocities and the CBF was found. The correlation between RI and CBF was poor.

In an animal experiment, Batton [35] compared Doppler flow velocities with the autoradiographically determined regional CBF during changes in the CO_2 partial pressure. For vs, vm, and vd the correlation with the regional

CBF was far better than for the RI. Rosenberg [332] obtained similar results in animal experiments on hypoxia.

Greisen [147] compared the Xenon-133 inhalation technique with continuous wave and pulsed Doppler sonography in newborns. With the Xe-133-CBF, vm and vd (r = 0.49–0.82) correlated better than RI (r = —0.41 to —0.56). The values measured with the pulsed method correlated more closely with the CBF than the findings of the continuous-wave method.

Hartmann [164] compared the results of the Xenon-133 inhalation technique with those of transcranial Doppler sonography in adults. The correlations found were highest under normocapnia and in healthy volunteers. The closest relationship was between Xe-133-CBF and vs (r = 0.77). The change in CBF when the CO_2 partial pressure changed was not clearly reflected in the blood flow velocities.

In the TCD investigations in healthy children reported here a close correlation between vs, vm, and vd and thus with SD and RI was demonstrated. Thus, vs and vm correlated more closely with each other than they each did with vd (Chapter 4.1.3., Appendices I–III).

Theoretical considerations

In infancy, the enddiastolic peak flow velocity (vd) is low. For this reason, the vd does not seem as well suited for differentiating the various biological states as is vs or vm. The vs, vm, and vd are independent of the heart rate, except in bradycardia in newborns, which makes them equivalent to a certain extent and thus preferable over the AUC. As an averaged velocity, vm has the advantage of being an expression of all peak flow velocities measured during a pulse wave.

A special significance of the velocity vm results from the following: multiplying the velocity vM (Chapter 2.4.4.), which is averaged over the entire cross-section of the vessel, with the cross-section of the vessel produces the flow volume (in ml/min) through the vessel examined (equation 3—Chapter 2.2.1.). In laminar flow, which is usually present in the large cerebral arteries [167], the mean axial = peak flow velocity vm is roughly twice the velocity vM [290]. The vm is calculated from the envelope curve of the Doppler waveforms. It is not necessary that the entire diameter of the vessel be in the sample volume [187]. Under the assumption that the diameter of the vessel is constant, the vm is directly proportional to the flow volume through the vessel. If the diameter of the vessel is known, the blood flow volume through the vessel can be quantitatively calculated from the vm [279, 384].

Significance of the vessel diameter

The diameter of the intracranial arteries cannot be measured free of artifacts by either sonography, angiography, or after death [215, 219, 344, 393].

The fluctuations of the vessel diameter dependent on cardiac cycle are minimal [274]. When the CO_2 partial pressure changes by 20 torr, the cross-sectional area of the distal ACA changes by about 30%, whereas that of the ICA changes by only 1% [171].

In the reported TCD measurements of flow velocities in the proximal basal arteries, the intraindividual diameter fluctuations probably played only a minor role in most cases [8, 399]. Greater lumen fluctuations are to be expected in the peripheral arteries. These arteries are muscle-type vessels [264] and regulate the cerebrovascular resistance.

Conclusions

In the intraindividual comparison it can be assumed—with the restrictions just mentioned—that increased flow velocities in the basal cerebral arteries are a sign of increased cerebral blood flow. Velocity changes thus indicate relative, but not absolute changes in the CBF [30]. The proportionality factor for the relationship between flow velocities and CBF is not constant.

The interindividual comparison is dealt with in Chapters 4.3. and 5.1.4.

5.1.4. Author's evaluation

Various Doppler parameters usually have to be taken into consideration when assessing the TCD-measured frequency spectra as a sign of the CBF, the diameter of the basal arteries examined, and the CVR. Frequently, it is necessary to state the vs, vm, vd, and in some cases the various indices [30, 147, 335, 399]. Furthermore, the analysis of the Doppler waveforms can be useful for special problems [6, 200]. The velocity increase in systole, for example, is influenced by the contractility of the cardiac muscle, by the heart rate, by the elasticity of the vessel, and by the viscosity of the blood. The velocity reduction in diastole is a sign of heart rate, compliance effect of the arteries supplying the brain, and the CVR.

In some diseases only single parameters of the Doppler spectra are normal or changed. Fig. 44 gives a summary of the results from Chapters 4.2. and 4.4. showing the Doppler recordings from the basal cerebral arteries under physiological (case 1) and pathological (cases 2–9) conditions.

Different diseases can produce similar Doppler waveforms. Raised flow velocities in the basal cerebral arteries can be an expression of an increased CBF (case 2) or of a stenosed basal cerebral artery (case 5). Doppler sonography of the extracranial brain-supplying arteries, which in infants and small children has not yet been standardized, could in the future help to decide which of the two is the case. It would demonstrate raised velocities in case 2 only. A markedly reduced arterial blood pressure or cardiac output (case 3) and a stenosis proximal to the examined artery (case 4) produce similar findings. The situation can be clarified by investigating several

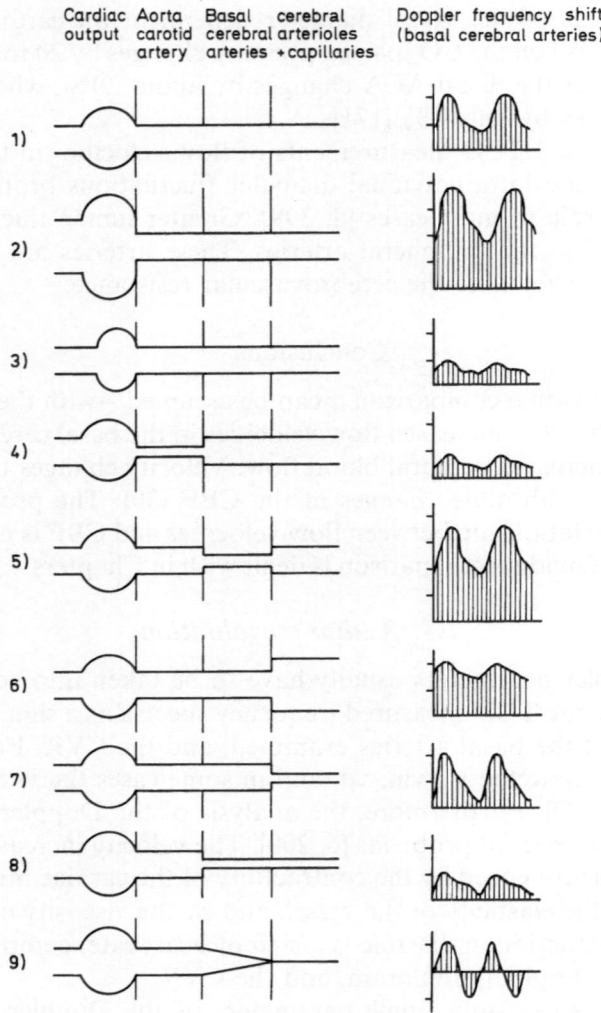

Cardiac   Aorta   Basal   cerebral   Doppler frequency shift
output   carotid cerebral arterioles   (basal cerebral arteries)
          artery   arteries +capillaries

1)
2)
3)
4)
5)
6)
7)
8)
9)

Fig. 44. Scheme of Doppler recordings from the basal cerebral arteries under physiological (*1*) and pathological (*2–9*) conditions

vessels on both sides. In the presence of hypercapnia and after cerebral hypoxia/ischemia the cerebral arterioles and capillaries become widened (case 6), while in hypocapnia and sometimes after extensive neonatal cerebral hemorrhage these vessels become narrow (case 7).

Although they have a different pathogenesis, a patent ductus arteriosus and increased intracranial pressure can produce similar Doppler waveforms in the basal cerebral arteries. In case 8 the signals are recorded in the presence of marked hypocapnia or increased intracranial pressure, whereas case 9 indicates brain death. Details will be discussed in the following sections.

Because the Doppler waveforms are not specific, the evaluation of the recordings should take into account the clinical condition of the child, the laboratory results, the results of imaging methods, and observations of the course of the disease. This applies particularly for recordings in the neonatal period.

5.2. Methods

5.2.1. Insonation angle

The Doppler shift is dependent on the real blood flow velocity, the emitting frequency of the Doppler instrument, the sound velocity in the tissue, and the angle between the sound beam and the axis of the blood flow (equation 10—Chapter 2.4.2.). The emitting frequency and the sound velocity are constant, the insonation angle not necessarily so. For this reason various authors have introduced Doppler parameters that are independent of the insonation angle. The resistance index RI and the pulsatility index PI have gained special importance [27, 144]. The Doppler instrument used calculates the SD ratio, which can be converted into the RI (Chapter 2.4.4.). It is independent of the angle, as is the SM ratio introduced for the diagnosis of the patent ductus arteriosus and the steepness of the rise and fall of the Doppler envelope curve [146].

On the basis of anatomic [219, 342], sonographic [191, 196], and CT examinations [160], it has been assumed that in investigations using transcranial Doppler sonography the angles between the sound beam and the axis of the blood flow are less than 30° [5, 6], giving measuring error for the blood flow velocities of below 13%. More recent studies [105a] using the TCD scanning system developed by Aaslid [6], which is more applicable to the TCD methodology, demonstrated averaged values between 9° and 17°, even in contralateral vessels. This is equivalent to an error between 1.1 and 4.4%. It can thus be expected that in repeated measurements on the same patient the same proportionality factor will exist between the measured Doppler shift and the real flow velocities in the vessel examined [30, 39, 399].

The variability of the course of the vessels [215, 219, 324, 393, 394] is a complicating factor in the interindividual comparison of the Doppler findings. A vessel is examined from the smallest angle possible when a probe position with optimal Doppler signals is considered.

5.2.2. Comparison of instruments

Doppler examination of the intracranial vessels was first carried out by means of the continuous-wave technique. Superimposing the Doppler signals from different arteries and veins, however, produced artifacts and hence false interpretations [8, 40]. The signal processing of these easy-to-use, inexpensive instruments resulted in an underestimation of the high flow velocities and an overestimation of the low flow velocities. The RI measured was lower than with the pulsed technique [147, 365, 387]. The intracranial vessels could only be examined through an open fontanelle.

Instruments using pulsed ultrasound produce more precise recordings, since they enable single vessels to be examined [8, 30]. Duplex-scan-sonography (DSS) and transcranial Doppler sonography (TCD) function according to the pulsed technique.

With both instruments, signal processing with frequency spectral analysis makes it possible to analyze the peak flow velocities. The sample volume of TCD (9×4 mm) is much greater than that of DSS (1.6×1.4 mm), which can have an influence on the recorded values. If the sample volume is too small, the axial = peak flow velocity or even the entire vessel to be examined can escape recording [399]. A large sample volume makes it easier to find a vessel without visual control and assures reliable recording of the peak flow velocity. The disadvantage, however, is that several vessels might be insonated at the same time. This particularly applies in newborns were the diameter of the ICA is 3–4 mm and that of the larger intracerebral arteries less than 1 mm [8]. It is therefore necessary to modify the TCD recording technique for application in infants (Chapter 3.3.).

Whereas recordings with DSS have to be made before the fontanelle closes, TCD can be carried out at any age. DSS allows visual control of the position of the sample volume; TCD does not (Chapters 3.1. and 5.2.3.). Like DSS, with TCD flow velocities of down to 1 cm/sec can be measured (Chapter 3.2.).

In comparing the values somewhat higher velocities were found with TCD than with DSS (Chapter 4.1.1.). There are probably several reasons for this:

1. The TCD recording of the axial = peak flow velocities is more reliable than that of the DSS

2. TCD found optimal Doppler signals in a more proximal segment of the ACA (pars ascendens) than did DSS (splenium corporis callosi)

3. The children were more alert during TCD, which was performed after DSS.

There is no major difference between TCD and DSS in terms of accuracy and reproducibility (Chapter 5.2.3.). With TCD continuous recording can be made (Chapter 6.1.). TCD instruments are much smaller, easier to operate, and less expensive than DSS devices.

These advantages have made TCD the method of choice for Doppler investigations of intracranial arteries in pediatrics for research, clinical diagnostics, and control of therapy. This will be substantiated in the following.

5.2.3. Intraindividual variability

Reproducibility

The intraindividual variation coefficient (V) of the flow velocities measured with TCD is determined by the actual measuring error and the biological fluctuations [21, 95, 250] during the period of the recording. TCD has a measuring accuracy of ± 2 cm/sec. Since vd is less than vm and vs, a greater variation coefficient results for vd than for vs and vm under the assumption of an absolute measuring error for all of the TCD recordings. The same applies to the low variation coefficient for RI.

Volpe [399] and Jorch [196] found only slight deviations between measurements at intervals of a few seconds. In healthy children in unaltered stages of vigilance, the author was not able to detect a significant fluctuation of flow velocities between intermittent recordings within 30 minutes.

The variation coefficients of 2–8% (vs, vm), 3–12% (vd), 1–4% (RI), and 3–7% (SD) in these TCD recordings were clearly lower than the CW-values of 5–25% [39, 147, 399] and somewhat lower than the variation coefficients obtained with DSS [147, 196, 399] of 7–10% (vs), 12–19% (vd), and 3–4% (RI).

The good reproducibility of the flow velocities recorded with TCD must not be taken for granted. Again, the position of the sample volume cannot be visually controlled. With TCD, similar variation coefficients have been found in adults as were found in children [417]. If the examiner has enough experience and correctly uses the recording technique, visual control of the sample volume position will not be necessary and the instrument does not have to be so technically sophisticated. The investigation time will be only slightly longer for the experienced examiner. It still remains to be seen whether a further development of TCD which allows a three-dimensional visualization of the intracranial vessels [6] will reduce the measuring error even more. Experience to date indicate that such an innovation does not seem necessary in pediatrics.

Differences between the two sides

The variation coefficients of the flow velocities in both ICA's determined with TCD in infants were 0–4%. This corresponds to the results of DSS [196]. No values have been published for the MCA, which could not be well recorded prior to the application of TCD. The same results were obtained for the MCA as for the ICA. There are no data on the ACA because with conventional techniques it is not possible to distinguish between the two vessels in infancy with a sufficient degree of reliability.

No data are available on the flow velocity differences in the corresponding basal cerebral arteries in children beyond infancy. For the ACA and the PCA higher variation coefficients were found (3–8%) than for the MCA and the ICA (0–4%). The reason for this could have been the different frequency of variations in the course of the vessels. Harders [161] obtained similar findings with TCD recordings in adults.

Correlations between blood flow velocities

This is thought to be the only study (Chapter 4.1.3.) on the relationships between the flow velocities in the different cerebral arteries. These investigations in children of all ages showed that the MCA is the most reliable vessel for TCD recordings. This can be explained by the recording technique and by the anatomical location of the arteries. The ICA in infants and the ACA in older children are further away from the Doppler probe than the MCA, which makes them more difficult to insonate. The ACA has more anatomical variations than the MCA [215, 219, 324, 344] and in older children is insonated at a less favorable angle.

Conclusions

It has been found that in many pediatric patients the recording technique can be simplified. In most cases, individual recordings are sufficient because they can be reproduced. The vs and the vm have proved to be the most reliable parameters. If the flow velocities in the various cerebral arteries can be assumed to undergo the same type of change, only one representative artery has to be examined. The MCA of one side would appear to be the vessel of choice. This simplifies the recording procedure, particularly in repeated examinations.

If regional or local blood flow disturbances cannot be ruled out, both MCA's, ICA's, ACA's, PCA 1's, and the BA must be recorded.

Additionally recording the SIPH and PCA 2 produced no further information.

5.3. Physiological influences on cerebral hemodynamics

Cerebral hemodynamics result from the interaction of many single factors. Some of these factors are accessible to the Doppler investigation (Chapter 2.2.2.).

The cerebral blood flow is the sum of the volume that flows through the four arteries supplying the brain. In many cases the Doppler sonographically measured flow velocities in a large basal cerebral artery can be regarded as a relative indicator of the CBF as was described in Chapters 5.1. and 5.2. The cerebrovascular resistance is represented by the resistance index RI. The cross-sectional area of a cerebral artery affects the flow velocities and the volume flow.

5.3.1. Position of head and body

Recordings during orthostasis tests in adults show that the flow velocity in the basal cerebral arteries [63a, 162] decreases temporarily and then spontaneously returns to normal after a few seconds. The time constant of the autoregulation in adults is 0–5.2 minutes [240, 250]. In newborns, blood pressure measurements also demonstrated a vasomotor reaction of the peripheral vascular system when the body was suddenly tilted [277].

In preterm and term newborns TCD produced no clear findings during orthostasis tests, which was probably due to changes in the state of vigilance (Chapter 4.2.1.). In premature newborns no difference was observed in flow velocities when the baby's head was lowered and raised. The difference in the hydrostatic pressure of approximately 3 cm water column between the two positions effects a difference in intracranial pressure of 20–30% [101, 110, 408] and of 3% in the mean arterial pressure. Compensation for the cerebral perfusion pressure reduction is probably achieved less by a reduced cerebrovascular resistance (= autoregulation) than by an increased blood pressure with a rise in the cerebral perfusion pressure, keeping the CBF constant.

Older children were in a supine position during the examination. In these children the difference in hydrostatic pressure between a lying and sitting position is greater than in infants, and compensation either by autoregulation or an increase in blood pressure cannot be automatically assumed.

In postmortem examinations, marked flow reductions in the carotid and vertebral arteries were measured when the head and neck were turned with the body in a prone position. From this it was concluded that extreme head positions with the baby being in a prone position could be a cause of sudden infant death [339]. It was not possible to confirm the findings in living premature newborns (Chapter 4.2.1.). With a constant vessel diameter being assumed, the position of the head and body had no effect on the blood flow in the arteries examined.

5.3.2. Various biological states

The cerebral circulation is regulated by how much oxygen and substrates the brain requires [250]. The CBF in newborns decreases after meals [90, 314]. The CBF in premature infants has been found to be higher while they are awake than when they are asleep [148].

In the intraindividual comparison with mostly constant vessel diameters the flow velocities are proportional to the CBF.

Higher flow velocities were found in newborns while they were awake (Appendix V) than in babies of the same age and weight during NREM sleep (Appendix IVa–IVc).

In NREM sleep the flow velocities in newborns were 20–30% lower than in REM sleep [84, 195, 314]. This difference could not be observed in premature newborns [148].

In preterm and term newborns it was found that the biological state had a marked influence on the cerebral blood flow (Fig. 8, Chapter 4.2.2.). Spontaneous movements, such as a reaction to being aroused, resulted in a temporary increase in the flow velocities. Irregular breathing or hyperventilation when crying affected the values considerably. After hyperventilation a velocity increase to more than the initial value followed as a sign of reactive hyperemia in response to the initially reduced flow velocities. Constant values were determined in NREM sleep. These findings are evidence that the cerebral blood flow is dependent on blood pressure, intracranial pressure, and CO_2 partial pressure and that the autoregulatory capacity is still relatively low.

In older children higher flow velocities were measured at the beginning of the examination. This is considered to be a result of increased CBF caused by anticipation or perhaps by hypoventilation.

5.3.3. Age

Age is known to have an influence on the parameters of the peripheral hemodynamics [205], which in turn affect the cerebral circulation. This monograph deals only with those intracranial factors that change with age insofar as they are relevant for the interpretation of the Doppler findings.

Only two studies are known to deal with the influence of age on the flow velocities in the basal cerebral arteries in children [146, 369]. Up to the 2nd day of life, a marked drop of the RI was measured, which was assessed as an indication of the previously raised cerebrovascular resistance. In the majority of children no continuous orthograde flow could be measured in the first hours of life. In the first 3 days of life the velocities increased linearly.

In adults a velocity reduction of 10–20% is found between the 20th and 65th year of age [21, 162, 167].

For the first time reference values for the flow velocities in the various large cerebral arteries in children of all ages using TCD are presented in this study (Chapter 4.3.). Age has a tremendous influence on the velocities. They increase linearly during the first two months, then increasingly slower until the maximum is reached at 5–6 years of age. At this point the values are roughly 3–4 times the initial value measured at birth, after which they drop almost linearly to about 70% of the peak value up to the age of 16.

In interpreting these findings it must be taken into account that the flow velocities are proportional to the volume flow and are inversely proportional to the square of the vessel radius (equation 3, Chapter 2.2.1.).

Quantitative data on the radius of the basal cerebral arteries in relation to age are hardly to be found, and it is almost impossible to obtain recordings that are free of artifacts [132, 219, 344, 393]. The vessel radius increases 1.5 times between the 28th and 40th weeks of gestation and by adulthood has tripled again [132, 219, 352]. At the age of 6 years, it has reached the dimensions of a young adult [393]. The brain weight also triples during the first 6 years of life [99].

From this relatively constant ratio of the cross-sectional area of the basal cerebral arteries and brain weight, one can derive a proportionality between cerebral blood flow and flow velocities during the first 6 years of life. The recorded velocities increase 3–4 times during this time, which corresponds to the wide range of values for cerebral blood flow [147, 202, 209]. After the 6th year of age, the CBF decreases at almost the same speed as the flow velocities, since the vessel radius and brain weight no longer change during this period. The decrease in flow velocities in later adulthood is probably due more to a reduced CBF [21] than to a growing vessel radius.

Using TCD, the CBF can be semiquantitatively determined when the vessel diameter in relation to age—though unknown in the individual case—is considered.

The younger and more premature the children, the longer and more often were extremely low vd values found (Table 5 in Chapter 4.2.3., Appendix IVc). The pulsatile portion of blood flow was particularly high in these children. The connection between age and RI during the first days of life [146, 369] was found to be continued throughout the entire first year (Appendix IVe). The reduction of the RI, and hence that of the SD, was most pronounced in the first weeks of life.

These findings can be interpreted as a sign of a higher cerebrovascular resistance (CVR) throughout the entire first year of life. The cerebral resistance vessels are assumed to constrict in the first days of life [146, 369] and the further drop of the CVR up to the end of the first year is probably a result of a doubling of the capillary density in this period [96, 97]. The further increase in capillary density up to adulthood to 1.5 times the value measured at one year does not seem to reduce the CVR any further. On the other hand, the medial layer of the cerebral arteries is markedly thicker in newborns, so that at this time it resembles the body arteries more than in later life. The amount of elastic fibres increases considerably between birth and age 1 [166]. This effects a higher elasticity (and lower rigidity) of the cerebral arteries and thus contributes to a higher CVR.

The reduction of the CVR in the first year of life can partially explain the increase in CBF during this period (equation 9 in Chapter 2.2.2.).

The increase in brain weight, neuronal differentiation, and thus cerebral functions [99] requires a greater amount of oxygen and substrates. This necessitates an increase in CBF in the first years of life [116, 250, 385, 409, 410].

5.3.4. Birth weight/gestational age

Jorch [196] showed that the flow velocities rise linearly in the large cerebral arteries with increasing weight of the newborn. What he did not take into account, however, is that the current weight of the child is influenced by birth weight and gestational age, as well by the current age.

In this study it was shown that in heavier newborn babies, on the average, flow velocities were higher on the first day of life and in the further neonatal period (Chapter 4.2.4.). This relationship was particularly pronounced for the vd and thus for the RI and the SD. The lower the birth weight, the longer and more often the vd was less than 6 cm/sec (Table 9, Chapter 5.4.3.). The birth weight and gestational age of the neonates correlated closely.

The CBF is correlated with the gestational age. This was quantitatively demonstrated in examinations of the CBF in premature infants [149, 230]. The higher flow velocities measured in more mature, that is heavier, newborns are a sign of a higher CBF.

The lower values for vd and thus also for RI and SD in children with lower birth weight/gestational age are regarded as a sign of higher cerebrovascular resistance. The reason for this, apart from the lower capillary density in the brain [96, 97], could be the small amount of elastic fibres in the cerebral arteries of preterm infants, which is responsible for the elasticity of these vessels resembling that of the body arteries more at this time than later [89, 166, 365].

However, birth weight probably has its own effect on the amount of CBF in the first days of life. In comparison with eutrophic children the weight of the brain in dystrophic newborns—although it is low for gestational age—is nevertheless high in relation to their birth weight [58]. The lower flow velocities in dystrophic newborns in comparison with eutrophic children indicates a lower CBF.

The fact that in the first 10 days of life the vs and vm were determined more by age than by the weight of the children shows the importance of the perinatal and postnatal adaptation processes (Appendix VI).

5.3.5. Hematocrit

According to the Hagen-Poiseuille law (Chapter 2.2.1.), the blood flow volume in the cerebral vessels is inversely proportional to the viscosity of the blood.

Thomas [381] found a reduced CBF in adults with polycythemia and gave the reason for this as being increased blood viscosity. When there was a slight rise of the hematocrit, the viscosity of the blood increased rapidly. Häggendal [153] observed a change in the CBF only when the viscosity was reduced. With the flow velocities being high enough, he found only slight viscosity-related changes in the CBF. This he considered to be an effect of an intact autoregulation of the CBF.

In this study much lower flow velocities were found in the neonatal period than later. During this time the hematocrit deviates from the norm frequently, as does the viscosity of the blood.

Neonatal polycythemia has important clinical implications [213, 335, 418]. Circulatory disturbances in the large and small cerebral vessels are made responsible for the temporary neurological symptoms (hyperviscosity syndrome [333]) and the high incidence of intracranial hemorrhages and cerebral infarctions [15, 53, 272, 418].

Rosenkranz [335] and Kolni [213] found reduced flow velocities and an elevated RI in polycythemic newborns, which returned to normal following a partial exchange transfusion with plasma. This led them to conclude that prior to the transfusion, the CBF was reduced and the cerebrovascular resistance raised.

A reciprocal connection between hematocrit and flow velocities/age in premature newborns is illustrated in Figs. 13 and 14. In a pair comparison it was shown that children with raised hematocrit had only a significantly lower vs, while the children with lower hematocrit had a higher vs, vm, and vd, and a lower RI in comparison with the control group.

These results suggest the value of considering hematocrit values that strongly deviate from the norm when evaluating the Doppler findings.

It is not known, however, whether the cross-section of the examined vessels was the same in the groups compared. It can thus be assumed, but not concluded, that the CBF in relation to the control group was lowered in children with polycythemia and raised in those with anemia.

It remains unclear as to whether the Doppler findings indicate insufficient cerebral perfusion. A change in the CBF could represent a normal adaptation process to the raised values for oxygen content and release from the blood in persons with polycythemia and lowered values in those with anemia.

Under these aspects an exchange transfusion with plasma in cases of polycythemia and a blood transfusion for anemia would only seem indicated if there were clinical symptoms in addition to the lowered and accelerated flow velocities.

The role of reduced flow velocities and lowered CBF in the pathogenesis of intracranial hemorrhage and cerebral infarctions which often accompany polycythemia can only be clarified by further examinations.

5.3.6. Arterial blood pressure

The flow velocity in a vessel rises with increasing perfusion pressure. The CBF increases with the cerebral perfusion pressure if the autoregulation does not provide for constant perfusion of the brain. The cerebral perfusion pressure is essentially determined by the arterial blood pressure (Chapter 2.2.).

In the first months of life, the flow velocities and the arterial blood pressure behave similarly:

1. In relation to age: the blood pressure already increases spontaneously in very small premature newborns within the first hours of life [276]. In newborns it increases very quickly in the first 2–3 weeks, and after the 6th week the increase slows down [7, 52, 105, 124, 227, 376].

2. In relation to weight: in the first days of life, newborns with higher birth weight have higher blood pressure values. The diastolic blood pressure in newborns is higher than in premature infants, both absolutely and relative to the systolic blood pressure [111, 124, 277, 388].

3. In relation to vigilance: the systolic blood pressure in sleeping infants is lower than that in babies which are awake [208, 376]. The blood pressure amplitude is lower in NREM sleep than in REM sleep [84].

4. The blood pressure values, like the flow velocities, show considerable intraindividual differences which are caused by the biological factors [135].

The increasing arterial blood pressure in the first months of life (and thus the elevated cerebral perfusion pressure) should cause an increase in the CBF, in view of the decreasing cerebral CVR at this age (Chapter 5.3.3.). This would be reflected in an increase in the flow velocities (Chapter 5.1.3.).

In a group of healthy neonates and premature newborns in whom flow velocities and arterial blood pressure were simultaneously measured, a parallel increase in the mean values with aging was detected (Fig. 15). However, in the analysis of the entire data, as well as of the individual case, no significant correlation was found between the simultaneously measured flow velocities and blood pressure.

It can be assumed that the increase in blood pressure (and thus the cerebral perfusion pressure) causes a rise in the flow velocities in the first month of life (Chapter 5.3.3.). This effect is apparently superimposed by the autoregulation of the cerebral circulation, which has also been reported by other authors in examinations performed in this age group [275, 295, 296].

5.3.7. Heart rate

In healthy preterm and term newborns with heart rates of greater than 120/min, there was no connection between the heart rate during the recording and the flow velocities (Chapter 4.2.3.). Consequently, it was not necessary to take the heart rate into account in determining the reference values.

In an examination of premature infants who presented with apnea and bradycardia, Perlman [306] discovered that the vd was not reduced until the heart rate was less than 120/min. The vs was reduced with a subsequently absent diastolic forwards flow if the heart rate was less than 80/min. The changes in the velocities were closely related to the arterial blood pressure levels [306]. This was interpreted as an indication that the autoregulation in these situations had been abolished. The same findings were obtained during prefinal respiratory cessation (Fig. 16).

In contrast, in older children with bradycardia of below 40/min no major change was found in the proportion of vs to vd. The drop of the velocities in diastole took place more slowly in accordance with the heart rate, which indicated good compliance of the brain-supplying arteries (Fig. 17).

Arrhythmias had a tremendous influence on the Doppler signals in the basal cerebral arteries (Fig. 18). This demonstrated the dependence of the cerebral hemodynamics on the cardiac output.

5.3.8. Bilirubin and phototherapy

The bilirubin concentration in serum is not known to have any influence on the cerebral hemodynamics [413]. In an examination of newborns with severe hyperbilirubinemia no significant deviation of the flow velocities from the reference values was observed (Chapter 4.2.8.).

Phototherapy is said to increase the incidence of patent ductus arteriosus in premature newborns [334]. Doppler sonography revealed no changes in the cerebral hemodynamics due to phototherapy. The increase in flow velocities during and after phototherapy (Chapter 4.2.8.) can be explained by the fact that the children were asleep during phototherapy and by the effect of age.

In animal experiments, however, it could be demonstrated that the extent of the bilirubin deposits in the subcortical area is strongly related to the CBF and is much higher in the presence of regionally increased CBF caused by hypercapnia [67, 68]. An increased CBF, which is thus a possible risk factor for kernicterus in premature infants, could be detected by transcranial Doppler.

5.3.9. CO$_2$ partial pressure

The CO$_2$ partial pressure (pCO$_2$) has a major effect on the CBF in the brain [159, 163, 250, 319]. The arterial, transcutaneous, and endexpiratory pCO$_2$ are found to be in close correlation [18, 159]. The relationship between the CBF and the pCO$_2$ is s-shaped. Between 20 and 70 mm Hg the CBF increases almost linearly with the pCO$_2$ and beyond this range there are only slight changes in the CBF [80, 163, 225, 226, 319]. In the physiological range the CBF increases by 0.9–1.8 ml/100 g brain tissue/min per mmHg

pCO_2. Hypercapnia has been shown to increase the CBF by 100–120%. In the presence of hypocapnia it can be reduced to 40–60% of the physiological value. The time constant given for the acute CO_2 effect is 1–4 minutes [250].

The changes in cerebrovascular resistance are caused by the changes in tone of the resistance vessels. Hypocapnia leads to vasoconstriction, hypercapnia to dilatation [226, 385].

During hypercapnia with pCO_2 pressures of 80–90 mmHg and maximum vasodilatation the autoregulation of the cerebral vessels is lost. In the presence of hypocapnia the threshold of the autoregulation is shifted to the lower blood pressure values [163, 386]. Furthermore, hypercapnia causes the arterial blood pressure to increase, which in turn has an effect on the cerebral hemodynamics [18, 62, 257].

A positive correlation between pCO_2 and flow velocities could be documented in several Doppler investigations in children and adults [17, 33, 35, 89, 164, 191, 257]. The relationship followed an exponential function which in the physiological range closely approximated a straight line [257]. There was a negative correlation to the RI consistent with the changes in the cerebrovascular resistance [17, 18, 34, 89].

These findings were confirmed in this study. A reduction of the vd was only observed during slight hypocapnia (Fig. 19), whereas during more severe hypocapnia the vs also decreased (Figs. 20 and 44). In the presence of hypercapnia the vd increased more than the vs (Fig. 20). The changes, however, were not specific. The Doppler signals obtained for patent ductus arteriosus Botalli were similar to those for hypocapnia and the findings after perinatal hypoxia were not unlike those for hypercapnia (Chapters 4.4. and 5.1.4.).

The CO_2 reactivity of the cerebral hemodynamics is of special interest. It can be expressed as the difference in flow velocities per mmHg pCO_2-change or as the change in percent in flow velocities per mmHg pCO_2 relative to the flow velocities at 40 mmHg pCO_2 [191, 412]. Mean values of 3.5–6% have been calculated, which were possibly influenced by age. However, there were considerable interindividual fluctuations [18, 89, 191, 257], (see Chapter 4.2.9.). In adults the CO_2 reactivity is markedly lower in the presence of severe vascular stenosis and arteriovenous malformations [167, 240, 412]. Reduced CO_2 reactivity was found in newborns with chronic hypercapnia and with a large patent ductus arteriosus Botalli (Chapters 4.2.9. and 4.4.1.).

The above results have the following implications:

1. For a correct interpretation of the Doppler findings it is necessary to know the pCO_2 during the Doppler recording.

2. If the pCO_2 is abnormal, the measured values have to be converted into those for a normal pCO_2 to prevent misinterpretations. However, due

to the fact that the CO_2 reactivity differs in each case, there is no universally valid correction factor.

3. The CO_2 reactivity of the flow velocities can be used as a diagnostic criterion for certain diseases.

4. The effect of a ventilation treatment—especially hyperventilation for the prevention of cerebral edema—on the CBF can be controlled by intermittent Doppler monitoring (Chapter 4.4.3.) or by continuous Doppler monitoring (Chapter 6.1.).

5.4. Comparison of measured and reference values with the literature

There are very few data in the literature concerning flow velocities and resistance index in the cerebral arteries of healthy volunteers. In no case was the group investigated large enough for normal values—in the strict sense of the word—to be obtained. Furthermore, the ultrasound frequencies and Doppler techniques employed varied, which limits the comparability of the results (Chapter 2.4.3.).

Several authors determined the RI in preterm and term newborns. The pulsed Doppler technique produced a RI of about 0.75 [92, 146, 196, 213, 260, 302, 365, 387], which is consistent with TCD measurements shown in Appendix IVe. A reduction of the RI with increasing gestational age has also been pointed out [365], which was also observed in these examinations (Chapter 4.2.4.).

Jorch [196] reported the following flow velocities measured in newborns (depending on body weight): ACA — vs = 20–24, vm = 10–20, vd = 5–12; ICA — vs = 38–50, vm = 20–27, vd = 8–14. In the ACA of newborns of different age and weight Deeg [92] found vs = 41 ± 12, vd = 10 ± 4; Greisen [147] found vd = 10.8 ± 6.0 (all measurements in cm/sec). These results obtained with the much more complicated Duplex scan technique correspond to the values measured with TCD (Appendix IV, V).

TCD in adults recorded a vm of 60 cm/sec as the mean of the norm in the MCA. There were substantial interindividual differences [1, 160]. The velocities measured in 10–16 year olds using TCD were 15–30% higher than those in the various basal cerebral arteries in adults [1, 6, 21, 160, 162]. The order of velocity reduction in children was the same as in adults [21, 161]: MCA-ICA-SIPH-ACA-BA-PCA.

The measured and reference values of this study correlate well with the results obtained using other Doppler sonographic techniques. TCD offers a reliable and easy method for determining the normal level of the flow velocities in the various basal cerebral arteries throughout the entire childhood period. These data are being reported for the first time. The fact that they are very closely age-related must be taken into account when evaluating Doppler recordings in children.

5.5. Clinical applications

5.5.1. Patent ductus arteriosus Botalli

The timely diagnosis and therapy of a patent ductus arteriosus Botalli (PDA) in newborns [43, 212, 232, 258] is crucial for the prognosis of the disorder [37, 362]. PDA therapy involves some risks [37, 258] and must therefore be monitored closely. Besides the clinical criteria such as heart murmurs, pulse quality, and the precordial palpation findings [108, 177, 218, 242, 253, 260, 301, 334, 392, 414, 415], the non-invasive procedures including the analysis of the arterial pulse wave [253], echocardiography [177, 212, 218, 242, 334, 424], and Doppler sonography are used in the diagnosis of a PDA. Doppler recordings are made on the aortal and pulmonary window of the PDA [88, 392], the descending aorta [192, 194, 242, 260, 349], the common carotid, the brachial, and the femoral [83, 218, 414, 415], and especially the anterior cerebral artery [83, 92, 107, 108, 192, 242, 260, 310].

The fact that there are a number of methods for diagnosing a PDA demonstrates that none of them is absolutely specific. This explains the varying incidence of patent ductus reported in the literature.

In a PDA requiring therapy the Doppler findings showed a retrograde blood flow in the diastole in vessels with high elasticity (body vessels) [83, 242, 260, 349]. In the vessels with low elasticity (brain vessels) a reduction of the vd to the point of reversed flow and accordingly an increased resistance index RI by 1 was found [92, 107, 108, 189, 242, 260, 301]. The data on the vs in the presence of a PDA varied [92, 192]. All of the Doppler examinations demonstrated a return of the flow velocities to normal after the PDA had been closed.

These findings were interpreted as a sign of a steal effect [242, 299] which the PDA had on the blood in the descending aorta [194] as well as in the large brain-supplying arteries. The changes were closely related to arterial blood pressure [301]. Speculations were made as to their significance for perinatal brain damage [37, 39, 192, 301].

The results with TCD in the presence of PDA (Chapter 4.4.1.) are consistent with the data reported in the literature. In this connection technical factors with regard to the particular instrument employed become evident. The TCD device used, for example, was not capable of measuring flow velocities below 6 cm/sec. In the case of PDA, however, the mean vd is between $+4$ and -6 cm/sec [92]. Consequently, it was only possible to estimate the vd and hence the RI. In no case was a reversal of flow observed in diastole. This disadvantage was largely compensated for by the additional criteria used to diagnose a PDA requiring treatment (Chapter 4.4.1.). The SM ratio (SM = vs/vm), unlike the RI, could always be calculated. The SM and RI values were proportional, since the vm was reduced in accordance with the vd. After they had reached their systolic maximum the flow velocities usually underwent a concave-shaped decrease, which was convex-shaped in the presence of a PDA (Fig. 22).

The TCD findings on intracranial arteries in the presence of a PDA in premature newborns revealed a disturbance in the proximal vascular bed which concealed the compliance effect of the brain-supplying arteries as well as the autoregulation of the cerebral vessels.

In general, the TCD findings were consistent with the clinical results (Table 11 in Chapter 4.4.1.). A PDA in need of treatment could be diagnosed by means of TCD. The more recent devices (Chapter 3.2.) that can register flow velocities of 1 cm/sec improve the results even more.

As a result of these investigations and theoretical considerations there are several objections to other studies:

1. In many studies, recordings on a comparable normal population are lacking. In Chapter 4.2. it was demonstrated that the level of flow velocities is strongly influenced by the child's age and birth weight. Low vd in small preterm newborns is the rule and not the exception so that this criterion is not sufficient for the diagnosis of a PDA that requires treatment. The importance of a normal population also becomes evident when assessing retrograde blood flow in the body arteries as a criterion for a patent ductus arteriosus [192, 260]. Other authors consider this finding normal [242, 414, 415].

2. Until now it has been ignored that the spontaneous or therapeutic closure of a PDA is a dynamic process and not an abrupt one [347, 362, 404]. This process was observed in several children with a large PDA. It is difficult to arrive at a definitive diagnosis on the basis of one single Doppler examination.

3. Most authors hardly ever state the CO_2 partial pressure in children with PDA at the time of the Doppler examination. Hypocapnia can cause a marked decrease of the vd, which can lead to a false-positive result. During hypercapnia, for instance in the case of a hyaline membrane disease, a false-negative finding can result.

4. Many reports neglect to give data on the incidence of other impairments common in children of this age and birth weight. Cerebral hemorrhage, intracranial pressure elevation, arterial hypotonia, and a truncus arteriosus communis can produce Doppler findings similar to those of a PDA.

These factors must be taken into account when examining the large cerebral arteries in preterm infants in order to diagnose a PDA reliably.

Beyond neonatal age, the PDA usually had no effect on the cerebral circulation in the patients studied. The examples of Blalock-Taussig shunt [350] (Fig. 23 in Chapter 4.4.1.) and subclavian steal syndrome [299, 321] demonstrate, however, that the changes in the cerebral hemodynamics typical for a PDA are not confined to preterm newborns in the first weeks of life.

In newborn infants with persistent fetal circulation, no effect on the Doppler signals was to be expected because the blood flow through the PDA was in the opposite direction.

5.5.2. Perinatal brain damage

In 1941, perinatal asphyxia was responsible for 50% of the deaths of newborn infants who had been born alive [76]. Today, it is still relevant for the mortality and developmental prognosis in an estimated 1.5–6 per 1,000 preterm and term infants born alive [152, 199, 238, 371, 398, 403].

Pathophysiology

Many factors are involved in the pathogenesis of perinatal brain damage [14, 77, 81, 85, 100, 121, 122, 125, 138, 140, 217, 221, 222, 235, 282, 289, 292, 294, 378, 379, 395, 397, 416, 421]. Apart from special morphological features of the cerebral vessels [76, 142, 165, 166, 251, 402], these include hypoxemia, ischemia, and impaired autoregulation of the cerebral circulation. It is not always possible to make a clear cut distinction between these conditions.

Hypoxemia: Prenatal or perinatal asphyxia can lead to hypoxemia and to hypoxia and lactate acidosis of the brain tissue. The acute hyperemia which sets in within a very short time causes brain swelling (hemodynamic swelling). The cerebral edema then develops in adults on the 2nd to 4th day and in children as soon as 12–24 hours after the damage has occurred [41, 42, 127, 139, 210]. Cerebral edema following asphyxia is probably the result of a disturbance of the blood brain barrier (=vasogenetic brain edema) as well as a cytotoxic component with cell hydrops [63, 115, 142, 210, 238, 286, 345, 346].

Ischemia: Asphyxia occuring in the fetus or newborn leads to arterial hypotonia and to global or regional cerebral ischemia in the area of the arterial watershed. The thromboembolic closure of a cerebral artery causes a focal cerebral ischemia [53, 243–246, 249, 251, 271].

Impaired autoregulation: Disturbed autoregulation of the cerebral blood flow is believed to be a pathophysiological mechanism involved in perinatal brain damage. This theory, however, is still a matter of controversy [9, 122, 243–246, 273, 386, 397]. In healthy animal fetuses and in preterm newborns, the range of autoregulation is limited. The blood pressure is close to the lower limit [295, 296, 383]. The autoregulation capacity is particularly vulnerable in the presence of hypercapnia, hypoxia, and acidosis [383]. A linear relationship between arterial blood pressure and the CBF was found in asphyxiated newborns, in preterm newborns with hyaline membrane disease, and in animal experiments. From this it was concluded that the autoregulation of the CBF had been abolished [9, 10, 244, 245]. Changes in the CBF that varied with region were found in the presence of pathological blood pressure levels [223, 297, 298].

Pathogenesis of perinatal brain damage: The periventricular region in the area of the germinal matrix in premature newborns is a vulnerable area for ischemia [22, 297, 422, 423]. Even mild arterial hypotonia can cause ischemic brain damage. In term infants ischemia is known to produce different patterns of damage. Cerebral ultrasound and computer tomography have shown characteristic morphological sequelae [19, 26, 53, 75, 175, 176, 234, 255, 256, 261, 291, 355, 357]. In many children the prognosis for development is poor [53, 122, 216, 246, 247, 371, 403].

If arterial blood pressure goes up spontaneously or as a result of external factors, the global and regional CBF increases when the autoregulation is impaired [298]. In severe cases the danger exists of capillary rupture with subsequent cerebral hemorrhage [155, 243, 395, 396]. Furthermore, premature newborns lack collagen and elastin in the walls of parenchymal cerebral arteries [165, 166], which increases the danger of these vessels rupturing when the intravasal pressure is elevated. In preterm newborns the favored site for cerebral hemorrhage is the rostral germinal layer [243].

Extensive ischemic lesions can accompany cerebral hemorrhage [36, 106, 269, 270, 357, 368, 401]. However, it has yet to be established [122] whether cerebral hemorrhages occur secondarily in a hypoxically damaged area [53, 291, 357, 371], or whether they cause ischemia due to vasoconstriction following subarachnoid hemorrhage, as is the case with adults [2–4, 11, 12, 162, 328, 329, 348].

Doppler findings

With Doppler sonography, changes in the cerebral hemodynamics associated with perinatal brain damage can be examined non-invasively. However, there are some limitations which must be taken into account:

— quantitative measurement of CBF is not possible (Chapter 5.1.4.).

— regional circulatory disturbances can only be detected in arteries accessible to Doppler examination (Fig. 26).

— multimorbidity and treatment strongly influence the results. It is difficult to trace the findings back to only one factor.

— the measured values may only be interpreted on the basis of reference values in terms of method, age, and weight. In many studies this has not been the case.

In order to record pathological findings qualitatively, the results of repeated measurements must be evaluated and the comparison of sides is mandatory, as is a knowledge of the reference values.

Cerebral edema after perinatal asphyxia: Bada [27] was the first to report Doppler findings of perinatal asphyxia. A reduced RI was considered a sign of dilated brain vessels [27, 367]. This was suggested as an attempt to restore the CBF, which was reduced as a result of a postasphyctic brain edema, by decreasing the cerebrovascular resistance [41]. A relationship between the duration and extent of the RI increase on the one hand and the prognosis of the children on the other was described [27, 41, 42].

The examinations in this study have shown that just a few hours after asphyxia, the vs and particularly the vd are already raised and the RI is reduced. However, at this point there is no indication of a true edema [115, 210]. It is therefore felt that the accelerated flow velocities are more a result of a maximum dilatation of the cerebral resistance vessels caused by local metabolites, by the opening of the otherwise closed collateral vessels, and by an increased CBF. These changes cause a hemodynamic swelling of the brain [76, 224, 245].

Thus, it is considered that the edema is not the cause [41, 42], but rather the result of the postasphyctic hyperemia and hyperfusion of the brain [142].

This concept can be used to justify the treatment of hyperventilation in asphyxiated newborns, the objective of which is to reduce the edema-promoting cerebral hyperperfusion (Chapter 5.3.9. and 5.5.3.). In the future Doppler sonography may be able to decide whether the asphyxia occurred before, during, or after birth.

The prognosis was unfavorable in the case of impaired autoregulation in mature newborns suffering perinatal asphyxia. Pressure passive changes in flow velocities (Fig. 25) and characteristic Doppler curves were found (Fig. 17). These turned into the typical reverberating pattern (Figs. 25 and 37), which is caused by the compliance of the large basal cerebral arteries (Chapter 5.5.3.).

Cerebral ischemia: It is still a matter of debate as to whether the Doppler findings in the presence of a PDA (Chapter 4.4.1.) indicate cerebral ischemia [39, 40, 92, 192, 399]. In newborns in shock, reduced flow velocities were measured and interpreted as a sign of decreased CBF [193].

To date, little attention has been paid to cerebral ischemia due to hypocapnia. This can be detected with Doppler (Figs. 19 and 36).

Reduced flow velocities on one side of the brain in connection with the corresponding sonographic findings are to be regarded as a sign of focal ischemia (cerebral infarction) [367] (Fig. 26 in Chapter 4.4.2.).

Cerebral hemorrhage: Abruptly rising blood pressures caused a substantial increase in the intracranial blood flow velocities. This was seen in suctioning [304], pneumothorax [173], seizures [303] (Fig. 34 in Chapter 4.4.3.), and patent ductus arteriosus (Chapter 5.5.1.) and was consistent with an increasing incidence of intraventricular hemorrhage. In preterm infants weighing less than 1,500 g a fluctuating cerebral blood flow velocity in accordance with the arterial blood pressures was associated with a high incidence of cerebral hemorrhage [305].

Within certain limits, the RI, that is, the cerebrovascular resistance, changed according to the CO_2 partial pressure in preterm newborns [89] (Chapter 4.2.9.). Hypercapnia further reduced the autoregulation of the CBF (Chapter 5.3.9.). It also produced an increase in flow velocities, that is, in the CBF (Chapter 4.2.9.).

In preterm newborns a higher blood flow was found on the affected side at the time of the intraventricular hemorrhage, whereas a few days later the blood flow was markedly reduced [268, 269, 401]. Animal experiments demonstrated a reduction of the regional blood flow after blood had been instilled into the ventricles [36]. After intraventricular hemorrhage the RI was higher than the norm [27, 368].

Using TCD it was possible to detect raised flow velocities on the day of the hemorrhage and clearly reduced velocities on the following days (Chapter 4.4.2.).

These findings are interpreted as follows: in preterm neonates the interaction of the increased CBF, reduced cerebrovascular resistance, and impaired autoregulation in the presence of elevated blood pressure, blood pressure peaks, or hypercapnia lead to cerebral hemorrhage. Cerebral hemorrhage causes an increase in the cerebrovascular resistance and a reduction of the CBF.

The findings of Doppler ultrasound are non-specific and thus do not allow a diagnosis of cerebral hemorrhage [302].

Clinical significance

The information obtained through Doppler examination of the cerebral arteries has had a tremendous impact on the therapeutic measures for seriously ill preterm and term newborns. They are partly responsible for the decreasing incidence of cerebral hemorrhage [380].

Doppler sonography provides a rational control of the therapy. A reduced and a raised cerebral blood flow, as well as altered cerebrovascular resistance can be qualitatively detected [246, 338] (Figs. 19, 24, 36).

The foregoing findings can be obtained with all of the modern Doppler ultrasound techniques. Apart from being easy to use, TCD can also be employed for continuous recording, which is an advantage in high-risk groups (Chapter 6). As opposed to duplex-scan sonography, TCD cannot provide a morphological visualization of perinatal brain damage. To properly interpret the TCD findings morphological results must be additionally obtained.

In the future, Doppler ultrasound could provide vital information on the pathogenesis and for the prophylaxis of intracerebral hemorrhage. It would be useful to check impaired autoregulation and fluctuating cerebral blood flow as risk factors in children in danger of suffering perinatal brain damage ([69], Chapter 6). This could be done by continuous recordings and would make it possible to control the effect of prophylactic measures on the cerebral hemodynamics [38, 73, 82, 102, 170, 179, 237, 249, 300, 336, 416]. Until then, the findings of Doppler sonography will only be a starting point in describing the hemodynamic processes during and after perinatal brain damage.

5.5.3. Increased intracranial pressure

Physiology

The physiological connections between cerebral hemodynamics and intracranial pressure (ICP) were described in Chapter 2.2.2. In the individual case raised intracranial flow velocities in the presence of increased fontanelle pressure were found. In the intraindividual comparison the flow velocities were found to be proportional to the CBF (Chapter 4.4.3.). The spontaneous fluctuations of the fontanelle pressure are considered to be a result of a changing cerebral blood volume (CBV), which could be caused by alterations—probably autonomous ones—in the CBF [21, 250].

An increase in the central venous pressure reduces the venous flow from the brain. The CBV and the ICP rise and the CBF drops temporarily. This explains the marked reduction of the flow velocities during the Valsalva test in the infants examined and is further evidence of the close relationship between the CBF and the flow velocities.

Pathophysiology

Because the skull does not stretch, intracranial space-occupying lesions cause the ICP to increase. The ICP first rises slowly, taking advantage of the extracerebral space in the skull, thereafter increasing more steeply [129, 130, 197, 211, 307]. Such phenomena are less pronounced as long as the cranial sutures are still open.

The effect of an increase in ICP on the blood flow in the brain under pathological conditions undergoes four phases [13, 41, 42, 66, 127, 250, 346, 354]. During these phases, pressure waves with very complex formation mechanisms can be observed [127–129, 211, 310].

Phase 1: As the ICP increases, the venous blood flow decreases for short time and hence the CBV expands. As a result the intracranial venous pressure rises until the original pressure difference has been reached. The CBV returns to the initial value. The CBF remains constant due to the reduced cerebrovascular resistance (= intact autoregulation).

Phase 2: The further ICP rise leads to increased CBV. With diminishing cerebral perfusion pressure (CPP) and increasing cerebrovascular resistance the CBF drops (= exhausted autoregulation). The Cushing reflex then sets into action, leading to an increase in the arterial blood pressure and thus the CPP, with a temporary rise of the CBF.

Phase 3: As the ICP rises still further, the CBF decreases again. The ICP follows the arterial mean pressure passively (= paralyzed autoregulation).

Phase 4: The ICP reaches the level of the arterial mean pressure. When the circulation in the brain ceases, brain death occurs.

Intracranial pressure measurement

ICP (and thus CPP measurements) along with clinical signs of increased pressure such as the Glasgow coma scale [129, 130, 178, 307], play a major role in the diagnosis and therapeutic control of raised intracranial pressure. In infancy, the ICP and CPP are much lower than later in life [101, 236, 308, 405, 408]. In practice, continuous recording of the ICP is indicated in newborns and infants when the CPP is below 30 mmHg in newborns and infants and when it is below 40–50 mmHg in older children and adults. Although children can tolerate lower values for some time better than adults [127, 129, 139, 211], a connection does exist between ICP and prognosis [28, 42, 139, 211, 220, 317, 364].

The disadvantages of continuously monitoring the ICP are the possible artifacts associated with the measurement of the fontanelle pressure and the necessary invasiveness of epidural pressure recordings in older children [127, 129, 130, 197, 211, 307]. The CBF cannot be determined because the cerebrovascular resistance is not known. The Doppler examination of the cerebral hemodynamics in the presence of increased ICP supplies additional information.

Doppler examinations in hydrocephalus

To date, most Doppler investigations of increased intracranial pressure have been in newborns and infants with hydrocephalus [13, 28, 93, 174]. The RI has been shown to be higher than the norm, but decreased substantially after decompression. The behavior of the flow velocities varied. The vs was found to be raised, constant, or diminished, the vd constant or reduced. The correlation of the Doppler parameters with the ICP was either poor or non-detectable [13, 174]. A better relationship to the size of the ventricles was established [174].

The measurements in this study showed similar, although less pronounced, changes in the RI. Somewhat lower values were found for vs and vd before compression than afterwards; for vm no significant difference could be detected. The Doppler signals were not able to provide any individual information on the extent of the ICP (Chapter 4.4.3.).

Various Doppler techniques were employed for the examinations mentioned above. The patients varied with regard to age, as well as type and phase of the hydrocephalus. The level of the ICP was not always given. There was not always a comparable group in terms of age and weight, nor was there any information on additional diseases. The specificity of the results is therefore questionable. The pathophysiological explanations of the findings, which assume an increased resistance [28, 174] or an increased compliance [13] of the cerebral arteries, do not apply to each individual case.

When the autoregulation is intact and the ICP increases, the cerebrovascular resistance decreases in order to ensure a constant CBF (phase 1 of the pressure increase). Examinations in adults have shown that up to an ICP of 50 mmHg the CBF diminishes only negligibly and that decompression causes no significant increase [156, 346, 353]. This might not apply so much to preterm infants or infants with extremely dilated ventricles.

The extent or the dynamics of the ventricular dilatation, or both, possibly have a greater influence on the Doppler findings than the level of the ICP (Chapter 5.5.5.). Marked ventricular dilatation, even when the ICP is normal, leads to a stretching of the artery and a reduction of its lumen and along with it the danger of ischemic brain damage [13, 174, 419].

The children examined in this study were not premature and, in contrast to the majority of cases given in the literature, had only slightly to moderately enlarged ventricles. Both factors could explain the relatively slight change in Doppler signals.

It is the opinion of the author that it is almost impossible to diagnose ICP in the individual infant with hydrocephalus on the basis of a single Doppler examination. At present, Doppler can at best be used for repeated examinations. Decompression should be performed before—or at least as soon as—Doppler detects the symptoms.

In the future, computer-assistance in analyzing the Doppler waveforms may produce more reliable information [6].

Case reports

The TCD findings were helpful in understanding the pathophysiology of various diseases with increased ICP (Chapter 4.4.3.).

The reduced flow velocities on the affected side in the presence of an unilateral cerebral edema (Fig. 30) and an unilateral subdural hematoma (Fig. 31) are considered to be a result of a moved vessel. The MCA on the diseased side was recorded from an obtuse angle. This caused a reduction of all velocities. An increase in the cerebrovascular resistance or a reduction of the CBF on the affected side would have to assume a lost autoregulation. The flow velocities on the affected side would have been higher if the basal cerebral arteries had been compressed.

Raised flow velocities indicated a raised CBF which alone (seizure—Fig. 34) or additionally (plexus papilloma—Fig. 35) caused an increased ICP. A linear relationship between flow velocities and blood pressure (Fig. 25) on the one hand and flow velocities and ICP (Fig. 33) on the other, was a sign of the prognostically poor situation of paralyzed autoregulation (third phase of cerebral pressure increase).

Control of therapy

Hyperventilation treatment to reduce increased ICP plays a vital role in the therapy of cerebral edema [130, 211, 238, 286, 346, 364].

Only some patients reveal spontaneous hyperventilation [127, 346]. In others the vasoconstrictive effect of hypocapnia is used to diminish the cerebral blood volume and the CBF [286]. If there is global vasoparalysis no effect can be expected. In the presence of primary ischemic disturbances and extreme hyperventilation, cerebral ischemia threatens to follow as a consequence of therapy.

TCD has greatly improved the control of hyperventilation therapy. With a slight vasoconstriction of the basal cerebral arteries taken into account, the flow velocities decrease somewhat less than the CBF during hypocapnia.

It was possible to control hyperventilation therapy after perinatal asphyxia and reanimation using the age-related reference values. Phases of insufficient blood supply, in particular, could be detected in good time (Fig. 36). Hyperventilation was forced in the case of pathologically raised flow velocities (Fig. 24).

TCD has made it possible to examine the cerebral hemodynamics during increased intracranial pressure in patients of all ages. This has considerably expanded the range of applications for Doppler sonography.

5.5.4. Brain death

All patients who showed a reverberating flow pattern (Figs. 25 and 37) in TCD were brain dead according to clinical and electrophysiological findings [66]. A reverberating flow pattern indicates that cerebral arteries have changed from a system of low resistance to one of high resistance [265].

During systole, the blood only flows into the widening proximal basal cerebral arteries, in diastole the flow is retrograde, and the effective blood flow ceases (fourth phase of ICP increase). In the infants examined this could be assumed on the basis of absent pulsations of intracerebral arteries in the B-scan.

TCD is less susceptible to disturbances than the electroencephalogram and provides evidence regarding the cerebral hemodynamics [133]. TCD is especially valuable in acute situations, as it can rule out or indicate the circulatory cessation quickly and easily. Since it is non-invasive, it is ideal for repeated examinations.

TCD should also be considered a real alternative to cerebral angiography in childhood [66, 104, 265]. For legal reasons, however, angiography still may remain mandatory before transplant surgery is initiated.

5.5.5. Cerebral malformations

In Chapter 5.5.3. the Doppler findings in the case of hydrocephalus due to cerebral malformations were discussed. Apart from this, TCD produced abnormal findings only in individual cases.

Hemimegalencephaly is a rare condition marked by an overgrowth of brain tissue in the abnormal hemisphere, though the brain vessels themselves are neither increased nor widened [206]. The first Doppler recordings in this disease are reported here. The results show lowered flow velocities on the affected side (Fig. 38), which could be the result of shifted basal cerebral arteries and an increased cerebrovascular resistance with reduced CBF. The velocities continued to drop on both sides in the repeated examinations, which is concluded to be due to a reduction of the CBF in the presence of progressive disease.

It is difficult to differentiate between an extreme hydrocephalus internus and a hydranencephaly in cerebral sonography [54]. Transcranial Doppler has made this possible (Fig. 39). In hydranencephaly only low blood flow at the base of the brain was found. In extreme hydrocephalus internus the largely normal velocities indicated the existence of good circulation in the cortex. An increase in the RI to 1, which has been described in patients with acute severe hydrocephalus occurring postnatally [174], was not found in the case examined which already had signs of ventricular enlargement with no further progression a few weeks before delivery.

5.5.6. Stenosis and occlusion of basal cerebral arteries

Localized lumen narrowings of the basal cerebral arteries can be temporary or permanent, and can increase. Repeated examinations with TCD are therefore particularly important.

Several diseases can be differentiated: vasospasm after subarachnoid hemorrhage; arterial stenosis and occlusion of vascular or thromboembolic genesis; vessel narrowings associated with bacterial meningitis. The changes occuring in the neonatal period have already been discussed in Chapter 5.5.2.

Vasospasm

Vasospasm of the basal cerebral arteries occurs in 70–80% of patients with subarachnoid hemorrhage [4, 11, 162, 348]. The vascular spasm begins on the 4th day after the hemorrhage, reaches its maximum on day 10–17, and subsides again until day 40. The vasospasm and the additional microcirculatory disturbances cause the CBF to drop, which can result in neurological deficits [11, 12, 24, 162, 328, 329]. It is important to identify vasospasm because it can be successfully treated with nimodipine, thus improving the patient's prognosis [11, 12, 23, 24, 74, 131, 373, 406]. This can be done using TCD, which is a noninvasive and easily repeatable technique, which does not have the risks associated with angiography.

A reduced cross-sectional area of a vessel leads to an acceleration of the blood flow (Chapter 2.2.), which TCD was able to detect in the MCA more clearly than in the ACA due to the collaterals [2, 4, 162]. The degree of flow acceleration correlates with the prognosis. Depending on the stage and course of the disease velocities between 140 and 250 cm/sec for the vm in the MCA were regarded as critical [4, 162, 348]. In some of the patients low frequency "musical murmurs" could be heard during systole, which had been caused by vessel wall vibration or by periodic flow turbulences [3].

In this study, vasospasm was diagnosed in two children. In one child, who had been operated on for subdural hematoma, angiography confirmed the findings (Fig. 40). The child with multiple intracerebral hemorrhages following traumatic brain injury (Fig. 41) was not examined by angiography. For differential diagnosis a raised CBF as the cause of increased flow velocities has to be considered in this case. However, the time span between the trauma and the examination, the varying participation of the cerebral arteries, and the neurological sequelae make this unlikely. Moreover, vasospasm is known to occur after severe brain trauma [214].

Children who suffer severe brain trauma should always be examined for vasospasm. Nimidopine therapy can be effective in reducing the cerebral edema and in improving the clinical outcome [214].

Vascular stenoses and occlusions

There are many causes of stenoses and occlusions of the basal cerebral arteries in children [59, 123, 183, 184, 343]. In adults with vascular stenoses usually caused by arteriosclerosis, the following Doppler results have been found: an increase in the flow velocities in the area of the stenosis; velocities lower than those on the healthy side and a damped waveform of the Doppler signals beyond the stenosis; raised velocities or reverse flow in the collaterals; and "musical murmurs" [2–4, 113, 160, 161, 239, 318, 323, 326, 327]. A reverse flow has been described distal to the occlusions if the anastomosis was adequate [160].

Such findings were obtained in Doppler recordings of vascular stenoses and occlusions in children (Chapter 5.6.). In two children with Moyamoya syndrome they provided the essential information for the diagnosis and follow up of the disease. In Moyamoya (Japanese for "fog") syndrome, a rare disease of unknown origin, the intima in the large basal cerebral arteries is thickened. This results in stenosis—usually bilateral—or occlusion of the supraclinoid segment of the ICA, the proximal ACA, the MCA, and less commonly the PCA and BA [59, 71, 330]. This disease, which is primarily defined by angiography, owes its name to the developing basal collateral system with its fog-like appearance [262, 377]. The natural course of moyamoya disease extends from a good prognosis to a rapid deterioration [59, 71]. The results of vascular stenoses are regional circulatory disturbances and infarctions [87, 126, 330].

Using TCD, the author was able for the first time to identify noninvasively multiple vascular stenoses in children with undetected Moyamoya disease. The severe focal findings in the electroencephalograms recorded at the same time demonstrated the functional significance of these stenoses. The progressive course of the disease was recognized and it was possible to control the patency and hemodynamic effectiveness of two extracranial-intracranial bypasses [56] (Fig. 42).

Bacterial meningitis

In angiography, inflammatory changes of the basal cerebral cisterns following severe forms of bacterial meningitis are observed. They cause stenoses of the ICA and MCA and a considerable deceleration of the blood circulation with subsequent damage to the cerebral parenchyma [91]. There are three phases of cerebral arteriopathy in bacterial meningitis: 1) vasospasm caused by humoral factors and inflammatory infiltrations of the basal cerebral vessels; 2) vasodilatation, a paralytic phenomenon owing to damage of the wall muscles; and 3) organic stenosis, the result of repair processes [182, 420]. It could be confirmed intraoperatively that the outer diameter of the ICA during bacterial meningitis was substantially reduced [162].

In three out of 12 children with bacterial meningitis TCD recordings revealed that the vm in the ICA and proximal MCA were 200 cm/sec, which is 3 times the norm for that age (Fig. 43—Chapter 4.4.7.). Since the flow velocities in the distal MCA were clearly reduced, the raised velocities were interpreted as a sign of a vascular stenosis and not of an increased blood flow. This was supported by the children's clinical condition, the normal fontanelle pressure during the measurements, the detection of ischemic or secondary hemorrhagic lesions of the cerebral parenchyma several days later, and the severe residual damage the children suffered. Angiographic confirmation was dispensed with because of the high risk involved.

Cerebral ischemia following local vascular stenosis must be regarded as an additional, serious damage factor in bacterial meningitis. It should be investigated as to whether a better prognosis can be achieved in such cases by the early application of a calcium antagonist.

TCD has proved to be an important method to non-invasively demonstrate or exclude stenoses and occlusions of the basal cerebral arteries as the cause of neurological disturbances in children. It is an useful method for selecting and timing angiography. If therapy is indicated, it can be initiated early enough and monitored by TCD.

5.5.7. Vasomotor headache and migraine

During vasomotor headache and migraine the cerebral circulation is impaired [50, 61, 185, 263, 356]. Recently, neuronal functional disturbances have been considered to be the cause of the focal symptoms and the altered regional circulation of the brain [228]. Pathological TCD findings can only be expected if the changes in the regional cerebral circulation are considerable and apply to the circulation area of a large basal cerebral artery. This study has shown that changes are not to be expected during the intervals between attacks. During acute phases of complicated migraine no differences were found in flow velocities in the MCA according to side; other authors found only slight differences [114]. There is still a lack of systematic Doppler sonographic examinations performed on larger numbers of patients.

5.5.8. Other diseases

In closing, findings in other diseases should be briefly mentioned. The reduced flow velocities in the cases of anorexia nervosa are seen as a sign of a reduced CBF in this condition of vita minima. The raised flow velocities in the presence of intracranial vascular murmurs cannot be definitively interpreted due to a lack of angiographic findings. As such noises occurred in all of the arteries examined, they are considered to be a sign of an increased CBF rather than of local turbulences.

The raised vd in the presence of sepsis and hemolytic uremic syndrome indicate a dilatation of the cerebral resistance vessels.

5.5.9. Therapy monitoring

The role of TCD in monitoring special therapeutic measures has been discussed for the individual diseases. In pediatrics the following aspects should be particularly emphasized:

1. Control of the therapeutic success in patent ductus arteriosus Botalli (Chapter 5.5.1.).

2. Monitoring of the cerebral hemodynamics of preterm and term newborns under mechanical ventilation (Chapter 5.5.2.).

3. The influence on the cerebral hemodynamics of measures reducing intracranial pressure (Chapter 5.5.3.).

4. Control of the therapeutic success for intracranial vascular stenoses and occlusions (Chapter 5.5.6.).

6. Outlook

Further pediatric applications of transcranial Doppler sonography (TCD) can be envisaged for the future.

6.1. Additional examination techniques

A valuable supplement to TCD would be the Doppler examination of the extracranial arteries supplying the brain. This would make it easier to ascribe raised intracranial blood flow velocities to a vascular stenosis or

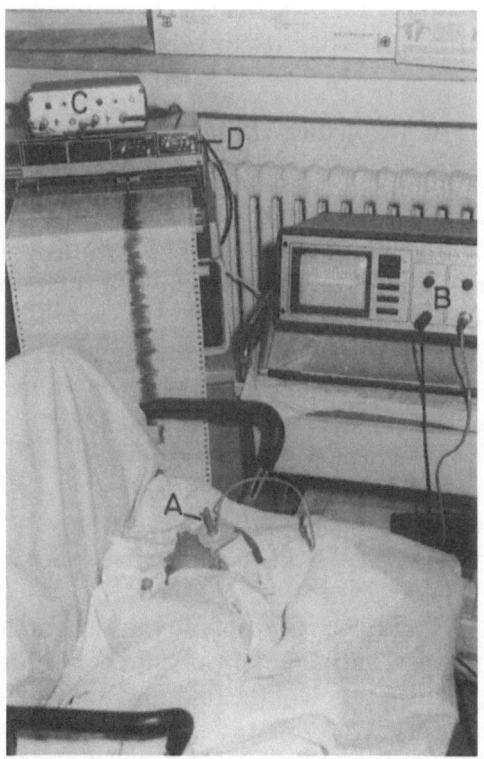

Fig. 45. Transcranial Doppler sonography—continuous recording technique: *A* Special holder and transducer, *B* Doppler device, *C* averager, *D* recorder

increased blood flow. To date, however, there have been no standardized recording techniques nor normal values for infancy and early childhood.

TCD of the large intracranial veins has never been systematically carried out. The method would be of particular interest in the case of perinatal brain damage and increased intracranial pressure.

The continuous recording of Doppler signals [69] would be advantageous in many clinical applications. A special holder has now been made available with which continuous recording is possible [109] (Fig. 45). In the first trials in newborns, there were problems associated with fixing the Doppler probe and processing the signals, which have since been overcome (unpublished results of the author). TCD will make it possible to monitor continuously the vital factor of cerebral hemodynamics, particularly in pediatric intensive care units. There are no known harmful effects of long-term insonation as long as the ultrasonic power used is below $40\,\mathrm{mW/cm^2}$ [331, 337].

Being able to determine the blood flow in the large arteries with TCD would be a great advance. This would necessitate determining the diameter of the vessel at the same time, which might be possible with ultrasound frequencies of more than $7\,\mathrm{MHz}$, provided these sufficiently penetrate the skull. This technical problem has yet to be solved.

6.2. Further clinical subjects

The use of TCD in pediatrics could enable various diseases and therapeutic procedures to be scientifically investigated and clinically monitored.

These would include: continuous TCD registration of
— preterm and term newborns at risk for perinatal brain damage
— comatose children
— treatment of hyperventilation therapy
— medication treatment (e.g. catecholamines, human albumin, and other plasma expanders, theophylline, muscle relaxants, barbiturates)
— heart surgery
— anesthesia
intermittent TCD recordings of
— altered blood viscosity (sickle cell anemia, spherocytosis)
— migraine
Simultaneous TCD and electroencephalogram examinations in patients in coma or with increased intracranial pressure could provide additional information for the interpretation of the findings. Simultaneous continuous recording of the Doppler waveforms and intracranial pressure might make it possible to determine the critical cerebral perfusion pressure and to estimate the hemodynamic importance of the pressure waves.

Comparative examinations of TCD and other, especially quantitative, methods for recording the cerebral hemodynamics (Chapter 5.1.) would be of interest for various diseases (perinatal asphyxia, increased intracranial pressure, brain trauma, brain death). The necessary invasiveness and possible side effects associated with these methods (Chapter 2.3.), however, will prohibit them from being systematically performed in children.

6.3. Doppler sonography and prognosis

The significance of Doppler findings of a lost autoregulation of the cerebral hemodynamics and of a narrowing of the basal cerebral arteries following brain trauma or bacterial meningitis for the prognosis seems dubious. The relationship between the Doppler spectra and prognosis that we established in individual cases has to be verified on the basis of a larger number of patients. It remains to be seen whether the Doppler findings in the neonatal period will provide information on the long term prognosis.

7. Appendices

Appendix I. Cross-sectional study on 50 healthy newborns between 1 and 10 days old (correlation coefficient matrix of the measured blood flow velocities)

	vsM	vmM	vdM	vsI	vmI	vdI	vsA	vmA	vdA
vsM	1	0.93	0.72	0.78	0.73	0.60	0.80	0.70	0.54
vmM		1	0.85	0.73	0.78	0.70	0.80	0.80	0.67
vdM			1	0.59	0.70	0.79	0.61	0.73	0.82
vsI				1	0.88	0.66	0.71	0.64	0.46
vmI					1	0.85	0.66	0.73	0.61
vdI						1	0.52	0.67	0.69
vsA							1	0.90	0.63
vmA								1	0.83
vdA									1

vs, vm, vd systolic, mean, enddiastolic peak flow velocity
M MCA, *I* ICA, *A* ACA
Thus, vsM means the systolic peak flow velocity in the MCA, etc. All correlations are significant with an error probability of $p < 0.001$.

Appendix II. Cross-sectional study on 40 healthy preterms older than 10 days of age. Correlation coefficient matrix of the measured blood flow velocities

	vsM	vmM	vdM	vsI	vmI	vdI	vsA	vmA	vdA
vsM	1	0.94	0.77	0.76	0.77	0.70	0.88	0.75	0.75
vmM		1	0.88	0.69	0.81	0.73	0.84	0.80	0.79
vdM			1	0.57	0.74	0.73	0.67	0.70	0.74
vsI				1	0.90	0.72	0.77	0.64	0.66
vmI					1	0.86	0.74	0.72	0.76
vdI						1	0.58	0.67	0.75
vsA							1	0.84	0.71
vmA								1	0.81
vdA									1

vs, vm, vd systolic, mean, enddiastolic peak flow velocity
M MCA, *I* ICA, *A* ACA
Thus, vsM means the systolic peak flow velocity in the MCA, etc. All correlations are significant with an error probability of $p < 0.001$.

Appendix III. Cross-sectional study on 112 healthy children of all age groups (correlation coefficient matrix of the measured blood flow velocities)

	vsM	vmM	vdM	vsI	vmI	vdI	vsA	vmA	vdA
vsM	1	0.97	0.94	0.97	0.95	0.92	0.88	0.85	0.87
vmM		1	0.97	0.95	0.97	0.95	0.87	0.86	0.89
vdM			1	0.92	0.95	0.96	0.82	0.81	0.88
vsI				1	0.97	0.93	0.88	0.88	0.87
vmI					1	0.97	0.85	0.88	0.89
vdI						1	0.82	0.84	0.87
vsA							1	0.97	0.94
vmA								1	0.97
vdA									1

vs, vm, vd systolic, mean, enddiastolic peak flow velocity
M MCA, *I* ICA, *A* ACA

Thus, vsM means the systolic peak flow velocity in the MCA, etc. All correlations are significant with an error probability of $p < 0.001$.

Appendix IV. Cross-sectional study on 112 healthy children of all age groups—dependence of blood flow velocities on age (mean value M and standard deviation S)
  * Examination through the anterior fontanelle
  + Examination from a high-frontal position

Appendix IV a. Systolic peak flow velocity (vs) in cm/sec

Age		MCA	ICA	SIPH	ACA	PCA 1	PCA 2	BA
		M/S	M/S	M/S	M/S	M/S	M/S	M/S
0–10	days	46/10	47/9*	—	35/8+	—	—	—
11–90	days	75/15	77/19*	—	58/15+	—	—	—
3–11.9	month	114/20	104/12*	—	77/15+	—	—	—
1–2.9	years	124/10	118/24	114/21	81/19	67/18	69/9	71/6
3–5.9	years	147/17	144/19	138/14	104/22	84/20	81/16	88/9
6–9.9	years	143/13	140/14	132/17	100/20	82/11	75/10	85/17
10–16.9	years	129/17	125/18	120/21	92/19	75/16	66/10	68/11

Appendix IVb. Mean peak flow velocity (vm) in cm/sec

Age		MCA	ICA	SIPH	ACA	PCA 1	PCA 2	BA
		M/S	M/S	M/S	M/S	M/S	M/S	M/S
0–10	days	24/7	25/6*	—	19/5[+]	—	—	—
11–90	days	42/10	43/12*	—	33/11[+]	—	—	—
3–11.9	months	74/14	67/10*	—	50/11[+]	—	—	—
1–2.9	years	85/10	81/8	75/10	55/13	50/17	50/12	51/6
3–5.9	years	94/10	93/9	91/11	71/15	56/13	48/11	58/6
6–9.9	years	97/9	93/9	89/11	65/13	57/9	51/9	58/9
10–16.9	years	81/11	79/12	77/14	56/14	50/10	45/9	46/8

Appendix IVc. Enddiastolic peak flow velocity (vd) in cm/sec

Age		MCA	ICA	SIPH	ACA	PCA 1	PCA 2	BA
		M/S	M/S	M/S	M/S	M/S	M/S	M/S
0–10	days	12/7	12/6*	—	10/6[+]	—	—	—
11–90	days	24/8	24/8*	—	19/9[+]	—	—	—
3–11.9	months	46/9	40/8*	—	33/7[+]	—	—	—
1–2.9	years	65/11	58/5	55/12	40/11	36/13	35/7	35/6
3–5.9	years	65/9	66/8	62/8	48/9	40/12	35/9	41/5
6–9.9	years	72/9	68/10	67/11	51/10	42/7	38/7	44/8
10–16.9	years	60/8	59/9	58/12	46/11	39/8	33/7	36/7

Appendix IVd. SD ratio (SD = vs/vd)—mean values

Age		MCA	ICA	SIPH	ACA	PCA 1	PCA 2	BA
0–10	days	4.8	4.8*	—	4.7[+]	—	—	—
11–90	days	3.4	3.4*	—	2.8[+]	—	—	—
3–11.9	months	2.6	3.0*	—	2.5[+]	—	—	—
1–2.9	years	2.2	2.1	2.3	2.2	2.2	2.1	2.2
3–5.9	years	2.5	2.5	2.7	2.3	2.4	2.4	2.5
6–9.9	years	2.3	2.2	2.2	2.3	2.2	2.1	2.2
10–16.9	years	2.4	2.4	2.4	2.4	2.2	2.3	2.3

Standard deviations:	0–10 days	:	2.2–2.8
	11–90 days	:	0.6–1.3
	3–11.9 months	:	0.4–0.6
	>1 year	:	0.1–0.4

Appendix IVe. Resistance index RI $= (vs-vd)/vs$—mean values

Age		MCA	ICA	SIPH	ACA	PCA 1	PCA 2	BAS
0–10	days	0.71	0.71*	—	0.64+	—	—	—
11–90	days	0.63	0.71*	—	0.60+	—	—	—
3–11.9	months	0.58	0.67*	—	0.60+	—	—	—
1–2.9	years	0.47	0.52	0.57	0.55	0.55	0.52	0.55
3–5.9	years	0.55	0.60	0.63	0.57	0.58	0.59	0.60
6–9.9	years	0.50	0.55	0.55	0.57	0.55	0.52	0.55
10–16.9	years	0.53	0.58	0.58	0.58	0.55	0.57	0.57

Standard deviations:	0–10 days	:	0.11
	11–90 days	:	0.07–0.10
	3–11.9 months	:	0.05–0.07
	>1 year	:	0.04–0.06

Appendix V. Increase in flow velocities (in cm/sec) with age—mean values M and standard deviations S from measurements taken from the longitudinal study on 24 healthy preterm and term newborns

	1st day M / S		5th day M / S		10th day M / S		15th day M / S		20th day M / S		25th day M / S	
vsM	42	10	54	10	61	13	64	10	70	13	80	12
vmM	22	6	26	6	30	6	32	6	37	7	41	8
vdM	13	4	15	4	16	3	17	5	19	6	20	4
vsI	35	6	44	10	51	8	52	8	62	8	60	8
vmI	18	5	22	6	27	5	28	4	33	5	34	6
vdI	11	4	12	3	14	3	14	3	16	3	15	3
vsA	32	7	42	7	44	8	52	7	57	5	60	11
vmA	17	5	21	6	23	5	27	5	31	3	33	7
vdA	11	5	13	3	13	3	16	5	16	2	18	3

vs, vm, vd systolic, mean, enddiastolic peak flow velocity
M MCA, *I* ICA, *A* ACA
Thus, vsM means the systolic peak flow velocity in the MCA, etc.

Appendix VI. Correlation coefficients for the relationships between flow velocities and age, birth weight, and gestational age

	Age		Birth weight	Gestational age
	Study CPN[a]	Study CP[a]	Study CPN[a]	Study CPN[a]
vsM	0.58	0.77	0.24	0.29
vmM	0.53	0.73	0.30	0.33
vdM	0.35	0.54	0.41	0.48
vsI	0.42	0.60	0.50	0.45
vmI	0.41	0.56	0.60	0.57
vdI	0.29	0.43	0.60	0.62
vsA	0.67	0.78	0.33	0.28
vmA	0.55	0.69	0.47	0.42
vdA	0.28	0.57	0.52	0.55

[a] Study CPN: cross-sectional study on 50 healthy preterm and term newborns between the ages of 1 day and 10 days. Linear relationships exist between birth weight and gestational age: $r = 0.86$.

Significances: $p < 0.05$ when $r > 0.282$
$p < 0.01$ when $r > 0.365$
$p < 0.001$ when $r > 0.456$

Study CP: Cross-sectional study on 40 healthy preterm newborns after the 10th day of life

Significances: $p < 0.05$ when $r > 0.325$
$p < 0.01$ when $r > 0.418$
$p < 0.001$ when $r > 0.519$

vs, vm, vd systolic, mean, enddiastolic peak flow velocity
M MCA, *I* ICA, *A* ACA
Thus, vsM means the systolic peak flow velocity in the MCA, etc.

Appendix VII. Constants (x 1—x 6) and correlation coefficients r of the regression curves for the flow velocities in the cross-sectional study on 112 healthy children. The correlation coefficient r designates the correlation of the measured value and estimated value.

Variable	x 1	x 2	x 3	x 4	x 5	x 6	r
vsM	13,468	0.0084	− 65.6	0.490	− 13,361	0.0086	0.92
vmM	12,372	0.0072	− 55.2	0.431	− 12,298	0.0073	0.94
vdM	7,382	0.0067	− 40.8	0.231	− 7,328	0.0068	0.94
vsI	13,467	0.0077	− 61.7	0.446	− 13,360	0.0078	0.91
vmI	12,787	0.0085	− 63.9	0.107	− 12,697	0.0086	0.92
vdI	10,701	0.0070	− 51.8	0.087	− 10,635	0.0071	0.93
vsA	11,138	0.0066	− 48.6	0.413	− 11,057	0.0067	0.83
vmA	9,372	0.0052	− 41.6	0.356	− 9,314	0.0052	0.83
vdA	1,820	0.0048	− 30.2	0.183	− 1,779	0.0049	0.86

vs, vm, vd systolic, mean, enddiastolic peak flow velocity
M MCA, *I* ICA, *A* ACA
Thus, vsM means the systolic peak flow velocity in the MCA, etc.

References

1. Aaslid R, Markwalder TM, Nornes H (1982) Noninvasive transcranial Doppler ultrasound recordings of flow velocity in basal cerebral arteries. J Neurosurg 57: 769–774
2. Aaslid R, Huber P, Nornes H (1984) Noninvasive transcranial Doppler ultrasound recording in basal cerebral arteries—a new approach to evaluation of cerebrovascular spasm. In: Voth D, Glees P (eds) Cerebral vascular spasm. W de Gruyter, Berlin New York, pp 287–294
3. Aaslid R, Nornes H (1984) Musical murmurs in human cerebral arteries after subarachnoid hemorrhage. J Neurosurg 60: 32–36
4. Aaslid R, Huber P, Nornes H (1984) Evaluation of cerebrovascular spasm with transcranial Doppler ultrasound. J Neurosurg 60: 37–41
5. Aaslid R (1986) The Doppler principle applied to measurement of blood flow velocity in cerebral arteries. In: Aaslid R (ed) Transcranial Doppler sonography. Springer, Wien New York
6. Aaslid R (1986) Transcranial Doppler examination techniques. In: Aaslid R (ed) Transcranial Doppler sonography. Springer, Wien New York
7. Adams FH, Landaw EM (1981) What are healthy blood pressures for children? Pediatrics 68: 268–270
8. Ahmann PA, Dykes FD, Lazzara A, Wilcox WD, Carrigan T (1983) Cerebral blood flow. Letter to the Editor. Pediatrics 71: 296–297
9. Ahmann PA, Dykes FD, Lazzara A, Holt PJ, Griddens DP, Carrigan TA (1983) Relationship between pressure passivity and subependymal/intraventricular hemorrhage as assessed by pulsed Doppler ultrasound. Pediatrics 72: 665–668
10. Ahmann PA, Dykes FD, Lazzara A, Giddens DP, Carrigan TA, Patino MM (1983) Cerebral blood flow characteristics in the ill premature: relationship to SEV/IVH. Pediatr Res 16: 331A
11. Allen GS et al. (1983) Cerebral arterial spasm-a controlled trial of nimodipine in patients with subarachnoid hemorrhage. N Engl J Med 308: 619–624
12. Allen GS (1983) Cerebral arterial spasm: A controlled trial of nimodipine in subarachnoid hemorrhage patients. Stroke 14: 122
13. Alvisi C, Cerisoli M, Ginlioni M, Monari PP, Salvioli GP, Sandri F, Lippi C, Bovicelli L, Pilu G (1985) Evaluation of cerebral blood flow changes by transfontanelle Doppler ultrasound in infantile hydrocephalus. Childs Nerv Syst 1: 244–247
14. Amiel-Tieson C (1981) Birth injury as a cause of brain dysfunction in full-term newborns. In: Korobkin R, Guilleminault C (eds) Advances in perinatal

neurology, vol I. SP Medical Scientific Books, New York London, pp 57–76

15. Amit M, Camfield PR (1980) Neonatal polycythemia causing multiple cerebral infarcts. Arch Neurol 37: 109–110

16. Anacker H, Allgayer B, von Einsiedel H, Reiser M, Rupp N, Halbsguth A, Lochner B, Graul EH (1985) NRM-Tomographie. DÄB 82: 2963–2970

17. Andrews AK, Samocha MS (1982) Effect of hyperventilation upon cerebral blood flow velocity in neonates with persistent pulmonary artery hypertension. Pediatr Res 16: 332A

18. Archer LNJ, Evans DH, Paton JY, Levene J (1986) Controlled hypercapnia and neonatal cerebral artery Doppler ultrasound waveforms. Pediatr Res 20: 218–221

19. Armstrong D, Norman M (1974) Periventricular leucomalacia in neonates. Complications and sequelae. Arch Dis Childh 49: 367–375

20. Arnold OH (1972) Störungen des Gehirnkreislaufs bei inneren Erkrankungen. In: Gänshirt H (Hrsg) Der Hirnkreislauf. G Thieme, Stuttgart, S 730–768

21. Arnolds BJ, von Reutern GM (1986) Transcranial Doppler sonography. Examination technique and normal reference values. Ultrasound Med Biol 12/2: 115–123

22. Ashwal S, Hewitt C, Longo LD, Loma L (1981) Regional cerebral blood flow studies in the fetal lamb during prolonged hypoxia, hypotension, acidosis and hypercapnia. Ann Neurol 10: 296A

23. Auer LM (1984) Acute operation and preventive nimodipine improve outcome in patients with ruptured cerebral aneurysms. Neurosurgery 15: 57–66

24. Auer LM (1985) Preventive nimodipine and acute aneurysm surgery. Heading for the control of complications after aneurysmal subarachnoid hemorrhage. Neurochirurgia 28 [Suppl] 1: 87–92

25. Babcock DS, Han BK, Le Quesue GW (1980) B-mode gray scale ultrasound of the head in the newborn and young infant. AJR 134: 457–468

26. Babcock DS, Ball W (1983) Postasphyxial encephalopathy in full-term infants: ultrasound diagnosis. Radiology 148: 417–423

27. Bada HS, Hajjar W, Chua C, Sumner DS (1979) Noninvasive diagnosis of neonatal asphyxia and intraventricular hemorrhage by Doppler ultrasound. J Pediatr 95: 775–779

28. Bada HS, Miller JE, Menke JA, Menten TG, Bashirn M, Binstadt D, Sumner DS, Khanna NN (1982) Intracranial pressure and cerebral arterial pulsatile flow measurements in neonatal intraventricular hemorrhage. J Pediatr 100: 291–296

29. Bada HS, Korones SB, Magill HL (1982) Alterations in cerebral hemodynamics in relation to onset of neonatal intraventricular hemorrhage. Pediatr Res 16: 332A

30. Bada HS, Sumner DS (1984) Transcutaneous Doppler ultrasound: Pulsatility index, mean flow velocity, end diastolic flow velocity and cerebral blood flow. J Pediatr 104: 395–397

31. Baird HW, Garfunkel JM (1953) A method for the measurement of cerebral blood flow in infants and children. J Pediatr 42: 570–575

32. Baker MD, Maisels MJ, Marks KH (1984) Indirect BP monitoring in the newborn. AJDC 138: 775–778
33. Batton DG, Hellmann J, Hernandez MJ (1982) Cerebral vascular resistance (CVR) and blood velocity in newborn puppies by Doppler technique. Pediatr Res 16: 332A
34. Batton DG, Hellmann J, Maisels MJ (1983) Doppler-pulsatility-index. Pediatrics 71: 298
35. Batton DG, Hellmann J, Hernandez MJ, Maisels MJ (1983) Regional cerebral blood flow, cerebral blood flow velocity and pulsatility index in newborn dogs. Pediatr Res 17: 908–912
36. Batton DG, Nardis EE, Hellmann J (1984) The effect of intraventricular hemorrhage (IVH) on regional blood flow in newborn dogs. Pediatr Res 18: 310A
37. Bedard MP, Shankaran S (1983) Closure of PDA re risk of IVH. J Pediatr 102: 1015
38. Bedard MP, Shankaran S, Slovis TL, Pantojy A, Dayal B, Poland RL (1984) Effect of prophylactic phenobarbital on intraventricular hemorrhage in high-risk-infants. Pediatrics 73: 435–439
39. Bejar R, Merritt TA Coen RW, Mannino F, Gluck L (1982) Pulsatility index, patent ductus arteriosus and brain damage. Pediatrics 69: 818–822
40. Bejar R, Merrit TA, Coen RW, Mannino F, Gluck L (1983) Doppler debate continued. Letter to the editor. Pediatrics 71: 471–472
41. van Bel F, Hirasing RA, Grimberg MTT (1984) Can perinatal asphyxia cause cerebral edema and effect cerebral blood flow velocity? Eur J Pediatr 142: 29–32
42. van Bel F, van de Bor M (1985) Cerebral edema caused by perinatal asphyxia. Helv Pediatr Acta 40: 361–369
43. Bell EF, Warburton D, Stonestreet BS, Oh W (1980) Effect of fluid administration on the development of symptomatic patent ductus arteriosus and congestive heart failure in premature infants. N Engl J Med 302: 598–604
44. Ben-Ora A, Eddy L, Hatch G, Solida B (1980) The anterior fontanelle as an acoustic window to the neonatal ventricular system. J Clin Ultrasound 8: 65–67
45. Berger H (1901) Zur Lehre von der Blutzirkulation in der Schädelhöhle des Menschen namentlich unter dem Einfluß von Medikamenten. G Fischer, Jena
46. Bergquivst G, Zetterström R (1974) Blood viscosity and peripheral circulation in newborn infants. Acta Paed Scand 63: 865–868
47. Bernmeier A, Gottstein U (1958) Hirndurchblutung und Alter. Verh Dtsch Ges Kreisl Forsch 24: 248
48. Betz E (1972) Pharmakologie des Gehirnkreislaufes. In: Gänshirt H (Hrsg) Der Hirnkreislauf. G Thieme, Stuttgart, S 411–433
49. Beverley DW, Chana G (1984) Cord blood gases, birth asphyxia and intraventricular haemorrhage. Arch Dis Child 59: 884–886
50. Blau JN (1978) Migraine: A vasomotor instability of the meningeal circulation. Lancet ii: 1136–1139
51. Bliesener JA (1981) Intrakranielle Veränderungen im Säuglings- und frühen

Kindesalter. Technik und Ergebnisse der Sonographie. Monatsschr Kinderheilkd 129: 200–215

52. Blumenthal S, Smith W, Tarazi RC, Lauer R, Jesse MJ (1980) Children's
    blood pressure in the United States. Pediatrics 66: 328–329

53. Bode H, Straßburg HM, Pringsheim W, Künzer W (1986) Cerebral infarction
    in term neonates: diagnosis by cerebral ultrasound. Child Nerv Syst 2: 195–
    199

54. Bode H, Straßburg HM (1986) Möglichkeiten einer sonographischen Dia-
    gnostik von Fehlbildungen des Zentralnervensystems beim Säugling. In:
    Neuhäuser G (Hrsg) Entwicklungsstörungen des Zentralnervensystems.
    Kohlhammer, Stuttgart

55. Bode H (1988) Transcranial Doppler sonography in infancy and early child-
    hood. In: Aaslid R, Eden A, Fieschi C, Zanette E (eds) Advances in trans-
    cranial Doppler sonography. Springer, Wien New York

56. Bode H, Sauer M, Harders A, Schumacher M (1987) Untersuchungen mit
    der transkraniellen Dopplersonographie bei einem Kind mit Moya-Moya-
    Erkrankung. In: Fichsel H (Hrsg) Aktuelle Neuropädiatrie 1986. Springer,
    Berlin Heidelberg New York Tokyo

57. Bradley J (1729) An account of a new discovered motion of the fixed stars.
    Phil Trans Roy Soc (London) 35: 637–661

58. Brand I (1981) Kopfumfang und Gehirnentwicklung. Klin Wochenschr 59:
    995–1007

59. Brett W (1983) Pediatric neurology. Churchill-Livingstone, Edinburgh Lon-
    don

60. O'Brien MJ, Ash JM, Gilday DL (1979) Radionuclide brain scanning in
    perinatal hypoxia-ischemia. Dev Med Child Neurol 21: 161–173

61. O'Brien MD (1971) Cerebral blood flow changes in migraine. Headache 10:
    139–143

62. Brubakk AM, Bratlid D, Oh W, Stonestreet BS (1984) Changes in epinephrine
    and mean arterial blood pressure during hypercarbia: effects on brain blood
    flow in the newborn piglet. Pediatr Res 18: 337A

63. Bruce DA, Goldberg A, Schut L (1977) ICP monitoring in critical care
    pediatrics. Int Care Med 3: 184–187

63a. Brunhölzl C, Müller HR (1986) Transkranielle Doppler-Sonographie in
    Orthostase. Ultraschall 7: 248–252

64. Büdingen HJ, von Reutern GM, Freund HJ (1982) Dopplersonographie der
    extrakraniellen Hirnarterien. Grundlagen-Methodik-Fehlermöglichkeiten-
    Ergebnisse. G Thieme, Stuttart New York

65. Büdingen HJ, Freund HJ (1985) Ultraschalldiagnostik an hirnversorgenden
    Arterien. Dtsch Ärztebl 82: 1843–1848

66. Bundesärztekammer (1986) Kriterien des Hirntodes. Dtsch Ärztebl 83: 2940–
    2946

67. Burgess GH, Oh W, Bratlid D, Brubakk AM, Cashore WJ, Stonestreet BS
    (1984) Cerebral hyperperfusion augments brain bilirubin deposition in pig-
    lets. Pediatr Res 18: 337A

68. Burgess GH, Oh W, Bratlid D, Brubakk AM, Cashore WJ, Stonestreet BS
    (1985) The effects of brain blood flow on brain bilirubin deposition in new-
    born piglets. Pediatr Res 19: 691–696

69. Busija DW, Heistad DD, Marcus ML (1981) Continous measurement of cerebral blood flow in anesthetized cats and dogs. Am J Physiol 241: H 228

70. Buy Ballot CHD (1845) Akustische Versuche auf der Niederländischen Eisenbahn nebst gelegentlichen Bemerkungen zur Theorie des Hrn. Prof. Doppler. Pogg Ann 66: 321–351

71. Cahan LD (1986) Failure of encephalo-duro-arterio-synangiosis procedure in Moyamoya disease. Pediatr Neurosci 12: 58–62

72. Chilowski C, Langevin MP (1916) Procédés et appareil pour production de signaux sous marins dirigés et pour la localisation à distances d'obstacles sous marins. French patent No. 502913

73. Chiswick ML, Johnson M, Woodhall C, Gowland M, Davies J, Touer N, Sims DG (1983) Protective effect of vitamin E (DL-alpha-tocopherol) against intraventricular haemorrhage in premature babies. Br Med J 287: 81–84

74. Cho C, Pruitt AW (1986) Therapeutic uses of calcium channel-blocking drugs in the young. AJDC 140: 360–366

75. Clancy R, Malin S, Laraque D, Baumgart S, Younkin D (1985) Focal motor seizures heralding stroke in full-term neonates. AJDC 139: 601–606

76. Clifford SH (1941) The effects of asphyxia on the newborn infant. J Pediatr 18: 567–578

77. Cole VA, Durbin GM, Olaffson A, Reynolds EOR, Rivers RPA, Smith JF (1974) Pathogenesis of intraventricular haemorrhage in newborn infants. Arch Dis Childh 49:722–728

78. Cooke RWI, Rolfe P, Howat P (1977) A technique for the non-invasive estimation of cerebral blood flow of the newborn infant. J Med Eng Technol 1: 263–266

79. Cooke RWI (1979) Ultrasound examination of neonatal heads. Lancet ii: 38

80. Cooke RWI, Rolfe P, Howatt P (1979) Apparent cerebral blood flow in newborns with respiratory disease. Dev Med Child Neurol 21: 154–160

81. Cooke RWI, Rolfe P (1979) Cerebral blood flow re intracranial haemorrhage. J Pediatr 95: 496

82. Cooke RWI, Morgan MEI (1984) Prophylactic ethamsylate for periventricular haemorrhage. Arch Dis Childh 59: 82–83

83. Couser RJ, Mammel MC, Coleman M, Boros SJ (1984) Continous wave Doppler velocity index: a new approach to the diagnosis of neonatal patent ductus arteriosus. Pediatr Res 18: 310A

84. Cowan F (1980) Blood pressure and cerebral blood flow in the healthy neonate. Arch Dis Childh 55: 159

85. deCrespigny LC, Mackay R, Murton LJ, Roy RND, Robinson PH (1982) Timing of neonatal cerebroventricular haemorrhages with ultrasound. Arch Dis Childh 57: 231–233

86. Cross KW, Dear PPRF, Warner RM, et al (1976) An attempt to measure cerebral blood flow in the new-born infant. J Physiol 260: 42–43

87. Dal-Bianco P, Reisner T, Friedrich MH, Spiel W (1983) Klinische und neuroradiologische Befunde bei infantiler Form der Moya-Moya-Erkrankung. Pädiatr Pädol 18: 29–37

88. Daniels O, Hopman JCW, Stoelinga GBA, Burch HJ, Peer PGM (1982)

Doppler flow characteristics in the main pulmonary artery and the LA/AO ratio before and after ductal closure in healthy newborns. Ped Cardiol 3: 99–104

89. Daven JR, Milstein JM, Guthrie RD (1983) Cerebral vascular resistance in premature infants. AJDC 137: 328–331

90. Dear PRF, Milligan DWA (1977) Measurement of cerebral blood flow in the newborn infant. Arch Dis Childh 52: 809

91. Decker K, Backmund H (1970) Pädiatrische Neuroradiologie. G Thieme, Stuttgart

92. Deeg KH, Gerstner R, Brandl U, Bundscherer F, Zeilinger G, Harai G, Singer H, Gutheil H (1986) Dopplersonographische Flußparameter in der Arteria cerebri anterior beim offenen Ductus arteriosus des Frühgeborenen im Vergleich zu einem gesunden Kontrollkollektiv. Klin Päd 198: 463–470

93. Deeg KH, Paul J, Rupprecht Th, Harms D, Mang C (1988) Gepulste Dopplersonographische Bestimmung absoluter Flußgeschwindigkeiten in der Arteria cerebri anterior bei Säuglingen mit Hydrozephalus im Vergleich zu einem gesunden Kontrollkollektiv. Mschr Kinderheilkd 136: 85–94

94. Deeply DT, Gordon RE, Hope PL, Parker D, Reynolds EOR, Shaw D, Whitehead MD (1982) Noninvasive investigation of cerebral ischemia by phosphorus nuclear magnetic resonance. Pediatrics 70: 310–313

95. Desmukh VD, Harper AM, Rowan JO, Jennett WB (1971/72) Studies on neurogenic control of the cerebral circulation. Europ Neurol 6: 166–174

96. Diemer K, Henn R (1964) The capillary density in the frontal lobe of mature and premature infants. Biol Neonate 7: 270–279

97. Diemer K (1964) Über die Gefäßversorgung des Gehirns im Säuglingsalter. Mschr Kinderheilkd 112: 240–242

98. Dittrich M, Straßburg HM, Dinkel E, Hackelöer BJ (1985) Zerebrale Ultraschalldiagnostik in Geburtshilfe und Pädiatrie. Springer, Berlin Heidelberg New York Tokyo

99. Dobbing J, Sands J (1973) Quantitative growth and development of human brain. Arch Dis Childh 48: 757–767

100. Dolfin T, Skidmore MB, Fong KW, Hoskins EM, Shennan AT (1983) Incidence, severity and timing of subependymal and intraventricular hemorrhage in preterm infants born in a perinatal unit as detected by serial real-time ultrasound. Pediatrics 71: 541–546

101. Donn SM, Philip AGS (1978) Early increase in intracranial pressure in preterm infants. Pediatrics 61: 904–907

102. Donn SM, Roloff DW, Goldstein GW (1981) Prevention of intraventricular hemorrhage in preterm infants by phenobarbitone. Lancet ii: 215–217

103. Doppler C (1842) Über das farbige Licht der Doppelsterne und einiger anderer Gestirne des Himmels. Abhdlg Königl Böhm Ges, Ser 2: 465–482

104. Drake B, Ashwal S, Schneider S (1986) Determination of cerebral death in the intensive care unit. Pediatrics 78: 107–112

105. Early K, Fayers P, Ng S, Shinebourne EA, deSwiet M (1980) Blood pressure in the first six weeks of life. Arch Dis Childh 55: 755–757

105a. Eden A, Whisler G (1988) Angle of insonation of the basal cerebral arteries through the transtemporal approach. In: Aaslid R, Eden A, Fieschi C,

Zanette E (eds) Advances in transcranial Doppler sonography. Springer, Wien New York

106. Editorial (1984) Ischemia and haemorrhage in the premature brain. Lancet ii: 847–848

107. Ellison P, Denny J (1981) Mean flow versus pulsatility index as an indicator of patent ductus arteriosus and intraventricular hemorrhage. Ann Neurol 10: 292

108. Ellison P, Eichhorst D, Rouse M, Heimler R, Denny J (1983) Changes in cerebral hemodynamics in preterm infants with and without patent ductus arteriosus. Acta Paediatr Scand [Suppl] 311: 23–27

109. EME, Eden Medizinische Elektronik GmbH (1985) TC 2–64, Transcranial Doppler. Operating instructions. Ueberlingen

110. Emery JR, Peabody JL (1983) Head position affects intracranial pressure in new born infants. J Pediatr 103: 950–953

111. Feldman RC (1982) Blood pressure in low-birth-weight infants. Pediatrics 69: 663–664

112. Feldtman RW, Andrassy RJ, Alexander JA, Stanford W (1976) Doppler ultrasonic flow detection as an adjunct in the diagnosis of patent ductus arteriosus in premature infants. J Thor Card Surg 72: 5–12

113. Ferbert A, Zeumer H, Ringelstein EB (1985) Dopplersonographische Befunde beim ischämischen Hirninfarkt unterschiedlicher Pathogenese. Akt Neurol 12: 153–157

114. Fiermonte G, Giacomini P, Sanarelli L, Pauri F, Pierelli F, Morocutti C (1986) TCD in patients with migraine and cluster headaches. First international conference on transcranial Doppler sonography, Rome, October 1986

115. Fishman RA (1975) Brain edema. New Engl J Med 273: 706–711

116. Flecknell PA, Wootton R, John M (1982) Accurate measurement of cerebral metabolism in the conscious, unrestrained neonatal piglet. II. Glucose and oxygen utilization. Biol Neonate 41: 221–226

117. Franklin DL, Schlegel WA, Rushmer RF (1961) Blood flow measured by Doppler frequency shift of backscattered ultrasound. Science 132: 564

118. Freeman J, Ingvar DH (1968) Elimination by hypoxia of cerebral blood flow autoregulation and EEG relationship. Exp Brain Res 5: 61

119. Freud HJ (1965) Ultraschallregistrierung der Pulsationen einzelner intrakranieller Arterien zur Diagnostik von Gefäßverschlüssen. Arch Psychiat Zschr Ges Neurol 207: 247–253

120. Freund HJ (1972) Ultraschallpulskurvenschreibung an Karotiden und Vertebralarterien. In: Gänshirt H (Hrsg) Der Hirnkreislauf. G Thieme, Stuttgart, S 392–401

121. Fricker HS, Sauter M, Buchs B (1983) Antepartale hypoxische Hirnschädigung (antepartale CTG- und postpartale CT-Veränderungen). Z Geburtsh Perinat 187: 50–53

122. Friis-Hansen B (1985) Perinatal brain injury and cerebral blood flow in newborn infants. Acta Paediatr Scand 74: 323–331

123. Fritsch G (1984) Acute infantile hemiplegia caused by cerebral infarction. Etiology, clinical features and investigations. Päd Pädol 19: 287–301

124. Fuchshofen M, Metze H (1976) Blutdruck von Frühgeborenen in den ersten drei Lebenswochen. Mschr Kinderheilkd 124: 596–598

125. Fujimura M, Salisbury DM, Robinson RO, Howat P, Emerson PM, Keeling JW, Tizard JPM (1979) Clinical events relating to intraventricular haemorrhage in the newborn. Arch Dis Childh 54: 409–414

126. Fukushima Y, Kondo Y, Kurok Y, Miyake S, Iwamoto H, Sekido K, Yamaguchi K (1986) Are Down syndrome patients predisposed to Moyamoya disease? Eur J Ped 144: 516–517

127. Gaab MR (1980) Die Registrierung des intrakraniellen Druckes. Grundlagen, Techniken, Ergebnisse, Möglichkeiten. Habilitationsschrift, Würzburg 1980

128. Gaab MR, Sörensen N, Hufenbeck B (1981) Fontanometrie zur nicht-invasiven Registrierung des intrakraniellen Druckes. Pädiatr Prax 24: 631–639

129. Gaab MR, Sörensen N, Bushe KA (1981) Überwachung des intrakraniellen Druckes mit gering invasiven Methoden beim schwerverletzten Kind. Techniken, Möglichkeiten und Ergebnisse. Z Kinderchir [Suppl] 33: 184–193

130. Gaab MR, Bushe KA (1981) Die Behandlung der intrakraniellen Drucksteigerung. Intensivbehdlg 6: 34–52

131. Gaab MR, Rode CP, Schakel EH, Haubitz J, Bockhorn J (1985) Zum Einfluß des Ca-Antagonisten Nimodipin auf die globale und regionale Hirndurchblutung. Klin Wochenschr 63: 8–15

132. Gabrielsen TO, Greitz T (1970) Normal size of the internal carotid, middle cerebral and anterior cerebral arteries. Acta Radiol Diagn 10: 1–10

133. Gänshirt H (Hrsg) (1972) Der Hirnkreislauf. G Thieme, Stuttgart

134. Geddes LA, Hoff HE (1964) The measurement of physiological events by electrical impedance-a review. Am J Med Electron 3: 16

135. Gillum RF (1980) Blood pressure study. Pediatrics 66: 1033

136. Gilsbach J (1983) Intraoperative Doppler sonography in neurosurgery. Springer, Wien New York

137. Gilsbach J (1984) Mikrovaskuläre intraoperative Doppler-Sonographie. Ultraschall 5: 246–254

138. Goddard-Finegold J, Armstrong D, Zeller RS (1982) Intraventricular hemorrhage following volume expansion after hypovolemic hypotension in the newborn beagle. J Pediatr 100: 796–799

139. Goitein KJ, Amit Y, Mussaffi H (1983) Intracranial pressure in central nervous system infections and cerebral ischaemia of infancy. Arch Dis Child 58: 184–186

140. Goldberg RN, Chung D, Goldmann SL, Bancalari E (1980) The association of rapid volume expansion and intraventricular haemorrhage in the preterm infant. J Pediatr 96: 1060–1063

141. Goldberg SJ (1984) A review of pediatric Doppler echocardiography. AJDC 138: 1003–1009

142. Goldstein GW (1979) Pathogenesis of brain edema and hemorrhage: role of the brain capillary. Pediatrics 64: 357–360

143. Gooding CA, Brosch RC, Lallemand DP, Wesby GE, Brant-Zawadzki MN (1984) Nuclear magnetic resonance imaging of the brain in children. J Pediatr 104: 509–515

144. Gosling RG, King DH (1974) Continous wave ultrasound as an alternative and complement to X-rays in vascular examinations. In: Reneman RE (ed) Cardiovascular applications of ultrasound. North-Holland, Amsterdam, pp 266–282

145. Graf M, Donis J, Sluga E (1986) Duplexscan der Carotiden: Flußwerte im Vergleich zu morphologischen Veränderungen. In: Otto RC, Schnaars P (Hrsg) Ultraschalldiagnostik 1985. G Thieme, Stuttgart

146. Gray PH, Griffin EA, Drumm JE, Fitzgerald DE, Duignan NM (1983) Continous wave Doppler ultrasound evaluation of cerebral blood flow in neonates. Arch Dis Childh 58: 677–681

147. Greisen G, Johansen K, Ellison P, Fredriksen PS, Mali J, Friis-Hansen B (1984) Cerebral blood flow in the newborn infant: Comparison of Doppler ultrasound and 133-Xenon clearence. J Pediatr 104: 411–418

148. Greisen G, Hellström-Vestas L, Lou H, Rosen J, Svennigsen N (1985) Sleep-waking shifts and cerebral blood flow in stable preterm infants. Pediatr Res 19: 1156–1159

149. Greisen G (1986) Cerebral blood flow in preterm infants during the first weeks of life. Acta Paediatr Scand 75: 43–51

150. Griffin D, Cohen-Overbeck T, Campbell S (1983) Fetal and uteroplacental blood flow. Clin Obstet Gynec 10: 565

151. Grolimund P (1986) Transmission of ultrasound through the bone. In: Aaslid R (ed) Transcranial Doppler sonography. Springer, Wien New York, pp 10–21

152. Haas G, Buchwald-Saal M, Mentzel H, Michaelis R (1983) Mortality and neurological morbidity in prematurely born infants and low birth weight infants born at term. Monatsschr Kinderheilkd 131: 733–735

153. Häggendal E, Norbäck B (1966) Effect of viscosity on cerebral blood flow. Acta Chir Scand [Suppl] 364: 13–22

154. Hales JRS (1974) Radioactive microsphere techniques for studies of the circulation. Clin Exp Pharm Physiol [Suppl] 1: 31

155. Hambleton G, Wigglesworth JS (1976) Origin of intraventricular hemorrhage in the preterm infant. Arch Dis Childh 51: 651–659

156. Hamer J, Stoeckel H, Alberti E, Weinhardt F (1973) Cerebral blood flow and metabolism in acute increase of intracranial pressure. Acta Neurochir 28: 95–110

157. Hansen NB, Stonestreet BS, Rosenkrantz TS, Oh W (1983) Validity of Doppler measurement of anterior cerebral blood flow velocity-correlation with blood flow in piglets. Pediatrics 72: 526–531

158. Hansen NB, Nowicki DT, Miller RR, Malone T, Bickers RG, Menke JA (1986) Alterations in cerebral blood flow and oxygen consumption during prolonged hypocarbia. Pediatr Res 20: 147–150

159. Hansen TN, Tooley W (1979) Skin surface carbon dioxide tension in sick infants. Pediatrics 64: 942–945

160. Harders A, Gilsbach J (1984) Transkranielle Doppler-Sonographie in der Neurochirurgie. Ultraschall 5: 237–245

161. Harders A, Gilsbach J (1985) Transcranial Doppler sonography and its application in extracranial-intracranial bypass surgery. Neurolog Res 7: 129–141

162. Harders A (1986) Neurosurgical applications of transcranial Doppler sonography. Springer, Wien New York

163. Harper AM, Glass HI (1965) Effect of alterations in the arterial carbon dioxide tension on the blood flow through the cerebral cortex at normal and low arterial blood pressures. J Neurol Neurosurg Psychiat 28: 449

164. Hartmann A, Ries F, Grolimund H, Tsuda Y (1986) Gefäßreaktivität, Blutfließgeschwindigkeit und Hirndurchblutung, gemessen mit transkranieller Dopplersonographie und der Xenon 133-Inhalationstechnik. In: Otto RC, Schnaars P (Hrsg) Ultraschalldiagnostik 1985. G Thieme, Stuttgart

165. Haruda F, Blanc WC (1981) The structure of intracerebral arteries in premature infants and the autoregulation of cerebral blood flow. Ann Neurol 10: 303A

166. Hassler O (1962) Elastic tissue contents of the medial layer of the cerebral arteries. Virch Arch Path Anat 335: 39–42

167. Hassler W (1987) Hemodynamic aspects of cerebral angiomas. Acta Neurochirurgica [Suppl] 37

168. Heiss WD, Biel C, Herholz K, Bewlik G, Wagner R, Wienhard K (1985) Atlas der Positronen-Emissionstomographie des Gehirns. Springer, Berlin Heidelberg New York Tokyo

169. Hennerici M, Daffertshofer M (1986) Multimodalitätsanalyse von Dopplersignalen der Carotis. In: Otto RC, Schnaars P (Hrsg) Ultraschalldiagnostik 1985, G Thieme, Stuttgart

170. Hensey OJ, Morgan MEI, Cooke RWI (1984) Tranexamic acid in the prevention of periventricular haemorrhage. Arch Dis Childh 59: 719–721

171. Hilal SK (1974) Cerebral hemodynamics assessed by angiography. In: Newton TH, Potts DG (eds) Radiology of the skull and brain. Mosby, St Louis

172. Hill A, Volpe JJ (1981) Measurement of intracranial pressure using the LADD intracranial pressure monitor. J Pediatr 98: 974–976

173. Hill A, Perlman J, Volpe JJ (1982) Relationship of pneumothorax to occurrence of intraventricular hemorrhage in the premature newborn. Pediatrics 69: 144–149

174. Hill A, Volpe JJ (1982) Decrease in pulsatile flow in the anterior cerebral arteries in infantile hydrocephalus. Pediatrics 69: 4–7

175. Hill A, Melson GL, Clark HB, Volpe JJ (1982) Hemorrhagic periventricular leucomalacia: Diagnosis by real-time ultrasound and correlation with autopsy findings. Pediatrics 69: 282–284

176. Hill A, Martin DJ, Daneman A, Fitz CR (1983) Focal ischemic cerebral injury in the newborn: diagnosis by ultrasound and correlation with computed tomographic scan. Pediatrics 71: 790–793

177. Hirschklau MJ, diSessa TG, Higgins CB, Friedman WF (1978) Echocardiographic diagnosis: Pitfalls in the premature infant with a large patent ductus arteriosus. J Pediatr 92: 474–477

178. Hofweber K (1982) Beurteilung einer Bewußtseinsstörung im Kindesalter mit Hilfe der Glasgow-Coma-Scale. Kinderarzt 13: 889–892

179. Hope PL, Thorburn RJ, Stewart AL, Reynolds EO (1981) Prevention of intraventricular hemorrhage by phenobarbitone. Lancet i: 527

180. Horbar JD, Yeager S, Philip A, Lucey JF (1980) Effect of application force on noninvasive measurements of intracranial pressure. Pediatrics 66: 455–457

181. Huber P (1972) Angiographische Funktionsdiagnostik des Hirnkreislaufs. In: Gänshirt H (Hrsg) Der Hirnkreislauf. G Thieme, Stuttgart

182. Igarashi M, Gilmartin RC, Gerald B, Wilburn F, Jabbour JT (1984) Cerebral arteriitis and bacterial meningitis. Arch Neurol 41: 531–535

183. Isler W (1969) Akute Hemiplegien und Hemisyndrome im Kindesalter. G Thieme, Stuttgart

184. Isler W (1973) Obstruction of intracranial arteries. Neuropädiatrie 4: 3–6

185. Jacobi G, Ritz A (1983) Migräne und vasomotorische Kopfschmerzen. Pädiatr Prax 28: 231–244

186. Jenkner FL (1962) Rheoencephalography: A method for the continous registration of cerebrovascular changes. Ch C Thomas, Springfield, Ill

187. Johnston KW, Maruzzo BC, Cobbold RSC (1977) Errors and artifacts of Doppler flowmeters and their solution. Arch Surg 112: 1335–1342

188. Joppich G, Schulte FJ (1968) Neurologie des Neugeborenen. Springer, Berlin Heidelberg New York, pp 296–308

189. Jorch G, Pfefferkorn J (1983) Direction of flow with PDA. J Pediatr 103: 833

190. Jorch G, Pfefferkorn J, Schneider W (1984) Flowmessung in der A. cerebri anterior mittels gepulster Doppler-Sonographie. Korrelation mit klinischen Faktoren. In: Kowalewski S (Hrsg) Pädiatrische Intensivmedizin VI. G Thieme, Stuttgart, S 36–39

191. Jorch G, Menge U (1985) Die Bedeutung des pCO_2 für die Hirndurchblutung in der Neonatologie. Monatsschr Kinderheilkd 133: 38–42

192. Jorch G, Reinhold P, Pollmann H, Pott U, Fischer U (1985) Gepulste Doppleruntersuchungen der Hirndurchblutung bei Frühgeborenen mit persistierendem Ductus arteriosus. Klin Pädiatr 197: 510A

193. Jorch G, Schneider W, Hentschel R, Ensterbrock T, Kurlemann G (1985) Gepulste Doppleruntersuchungen der Hirndurchblutung im neonatalen Schock. Klin Pädiatr 197: 510A

194. Jorch G, Matthäi I, Raupp U, Poland M, Ulrich K (1985) Dopplersonographischer retrograder flow in der A. coeliaca bei Frühgeborenen mit großem Ductus arteriosus und nekrotisierender Enterocolitis. Klin Pädiatr 197: 510A

195. Jorch G, Huster T (1986) State dependent changes of blood flow velocity in the anterior cerebral artery of neonates measured by pulsed Doppler investigations. Eur J Pediatr 144: 530A

196. Jorch G, Pfannschmidt J, Rabe H (1986) Die nicht invasive Untersuchung der intrazerebralen Zirkulation bei Früh- und Neugeborenen mit der gepulsten Dopplersonographie. Monatsschr Kinderheilkd 134: 804–807

197. Kaiser G, Pfenninger J (1984) Effect of neurointensive care upon outcome followings severe head injuries in childhood—a preliminary report. Neuropediatrics 15: 68–75

198. Kaneko J, Shiraishi J, Omizo H (1970) Analysis of ultrasonic blood rheogram by the sound spectrograph. Jap Circ 34: 1035–1045

199. Karch D (1984) Die Erkennung von zerebralen Funktionsstörungen und späteren Zerebralschäden bei Neugeborenen unter Intensivpflege. Klin Pädiatr 196: 336–341

200. Karickhoff A (1981) Simultaneous blood pressure and anterior cerebral artery flow velocity wave form analysis in the preterm infant. Pediatr Res 15: 666 A

201. Keller HM, Meier WE, Anliker M, et al (1976) Noninvasive measurement of velocity profiles and blood flow in the common carotid artery by pulsed Doppler ultrasound. Stroke 7: 370

202. Kennedy C, Sokoloff L (1957) An adaption of nitrous oxide method to the study of the cerebral circulation in children; normal values for cerebral blood flow and metabolic rate in childhood. J Clin Invest 36: 1130

203. Kennedy C, Grave GD, Jehle JW, Sokoloff L (1972) Changes in blood flow in the component structures of the dog brain during postnatal maturation. J Neurochem 19: 2423

204. Kety SS, Schmidt CF (1945) The determination of cerebral blood flow in man by the use of nitrous oxide in low concentrations. Am J Physiol 143: 53–66

205. Keuth U, Peusquens M (1956) Die hämodynamischen Kreislaufgrößen im Säuglings- und Kindesalter (Normalwerte und vergleichende Untersuchungen). Zschr Kinderhlkd 78: 379–400

206. King M, Stephenson JBP, Ziervogel M, Doyle D, Galbraith S (1985) Hemimegalencephaly-a case for hemispherectomy? Neuropediatrics 16: 46–55

207. Kisslo J, Adams D, Mark DB (eds) (1985) Basic Doppler echocardiography—The Doppler standards and nomenclature comittee of the American Society of Echocardiography: Recommendations for terminology and display for Doppler echocardiography. Churchill Livingstone, New York

208. Kirkland RT, Kirkland JL (1972) Systolic blood pressure measurement in the newborn infant with the transcutaneous Doppler method. J Pediatr 80: 52–56

209. Kirsch JR, Traystman RJ, Rogers MC (1985) Cerebral blood flow measurement techniques in infants and children. Pediatrics 75: 887–895

210. Klatzo I, Seitelberger F (eds) (1967) Brain edema. Springer, Wien New York

211. Klöti J (1985) Intrakranielle Druckmessung in der Pädiatrie-Technik, Indikationen und Therapiemöglichkeiten. Erg Inn Med Kinderheilkd 54: 1–33

212. Knight DB, Yu VYH (1986) Contrast echocardiographic assessment of the neonatal ductus arteriosus. Arch Dis Childh 61: 484–488

213. Kolni HW, Bada H, Korones SB, Fritsch CW, Ford DL (1982) Cerebral arterial pulsatile flow changes in neonatal polycythemia and hyperviscosity. Pediatr Res 16: 295 A

214. Kostron H, Twerdy K, Stampfl G, Mohsenipour I, Fischer J: Treatment of the traumatic cerebral vasospasm with the calcium channel blocker nimodipine: a preliminary report. Neurol Res 6: 29–32

215. Krayenbühl H, Yaşargil MG (1972) Das normale Hirngefäßsystem im an-
     giographischen Bild. In: Gänshirt H (Hrsg) Der Hirnkreislauf. G Thieme,
     Stuttgart, S 161–182
216. Krishnamoorthy KS, Shannon DC, deLong GR, Todres LD, Davies KR
     (1979) Neurologic sequelae in the survivors of neonatal intraventricular hem-
     orrhage. Pediatrics 64: 233–237
217. Künzer W, Pringsheim W, Niederhoff H, Hendrich-Schäfer G, Sutor AH
     (1977) Hirnblutungen beim Atemnotsyndrom. In: Mietens C (Hrsg) Das
     Atemnotsyndrom des Neugeborenen: Pathophysiologie, Therapie, Prognose.
     G Thieme, Stuttgart, S 61–75
218. Kupferschmid C, Lang D, Pohlandt F (1986) Vergleich der Befunde aus
     Doppler-Flußanalyse und klinischen und echocardiographischen Befunden
     bei Frühgeborenen mit symptomatischem Ductus arteriosus. 35. Jahresta-
     gung der Süddeutschen Gesellschaft für Kinderheilkunde, Augsburg
219. Lang J (1981) Klinische Anatomie des Kopfes. Neurokranium, Orbita, kra-
     niozervikaler Übergang. Springer, Berlin Heidelberg New York
220. Langfitt THW, Obrist WD, Gunnarelli TA, O'Connor MJ, ter Weeme CA
     (1977) Correlation of cerebral blood flow with outcome in head injuried
     patients. Ann Surg 186: 411–414
221. Laptook AR, Stonestreet BS, Oh W (1982) Regional brain blood flow and
     O_2-delivery during hemorrhagic hypotension in the piglet. Pediatr Res 16:
     295A
222. Laptook AR, Stonestreet BS, Oh W (1982) The effects of different rates of
     plasmanate infusions upon brain blood flow after asphyxia and hypotension
     in newborn piglets. J Pediatr 100: 791–796
223. Laptook AR, Stonestreet BS, Oh W (1983) Brain blood flow and O_2-delivery
     during hemorrhagic hypotension in the piglet. Pediatr Res 17: 77–80
224. Lassen NA (1966) Luxury perfusion. Lancet ii: 1113
225. Lassen NA, Ingvar DH (1972) Quantitative und regionale Messung der
     Hirndurchblutung. In: Gänshirt H (Hrsg) Der Hirnkreislauf. G Thieme,
     Stuttgart, S 342–349
226. Lassen NA (1974) Control of cerebral circulation in health and disease. Circ
     Res 34: 749
227. Lauer RM, Burns TL, Clarke WR (1985) Assessing children's blood pressure-
     consideration of age and body size: the Muscatine study. Pediatrics 75: 1081–
     1090
228. Lauritzen M, Olsen TS, Lassen NA, Paulson OB (1983) Changes in regional
     cerebral blood flow during the course of classic migraine attacks. Ann Neurol
     13: 633–641
229. Lautenbur PC (1973) Image formation by induced local interactions: ex-
     amples employing nuclear magnetic resonance. Nature 242: 190
230. Leahy FAN, Sankaran K, Cates D, Mac Callum M, Rigatto H (1979) Quan-
     titative noninvasive method to measure cerebral blood flow in newborn
     infants. Pediatrics 64: 277–282
231. Lechner H (1972) Impedanzmethoden. In: Gänshirt H (Hrsg) Der Hirn-
     kreislauf. G Thieme, Stuttgart
232. Leffler CW, Busija DW, Fletcher AM, Beasley DG, Hessler JR, Green RS

(1985) Effects of indomethacin upon cerebral hemodynamics of newborn pigs. Pediatr Res 19: 1160–1164

233. Lemburg P, Bretschneider A, Storm W (1981) Ultraschall zur Diagnostik morphologischer Hirnveränderungen bei Neugeborenen. Diagnostischer Wert der B-Bild-Methode. Monatsschr Kinderheilkd 129: 190–199

234. Levene MI, Wigglesworth JS, Dubowitz V (1983) Hemorrhagic periventricular leucomalacia in the neonate: a real-time ultrasound study. Pediatrics 71: 794–797

235. Levene MI (1983) Aetiological aspects and their implications in the prevention of neonatal intracranial hemorrhage. Eur J Obst Gyn Repro Biol 15: 335

236. Levene MI, Evans DH (1983) Continous measurement of subarachnoid pressure in the severely asphyxiated newborn. Arch Dis Childh 58: 1013–1015

237. Levene MI (1983) Protective effect of vitamin E against intraventricular haemorrhage in premature babies. Br Med J 287: 617

238. Levene MI, Evans DH (1985) Medical management of raised intracranial pressure after severe birth asphyxia. Arch Dis Childh 60: 12–16

239. Lindegaard KF, Bakke SJ, Grolimund P, Aaslid R, Huber P, Nornes H (1985) Assessment of intracranial hemodynamics in carotid artery disease by transcranial Doppler ultrasound. J Neurosurg 63: 890–898

240. Lindegaard KF, Aaslid R, Nornes H (1986) Cerebral arteriovenous malformations. In: Aaslid R (ed) Transcranial Doppler sonography. Springer, Wien New York

241. Linderkamp O, Strohhacker I, Versmold HT, Klose H, Riegel KP, Betke K (1978) Peripheral circulation in the newborn: interaction of peripheral blood flow, blood pressure, blood volume and blood viscosity. Eur J Pediatr 129: 73–81

242. Lipman B, Serwer GA, Brazy JE (1982) Abnormal cerebral hemodynamics in preterm infants with patent ductus arteriosus. Pediatrics 69: 778–781

243. Lou HC, Lassen NA, Friis-Hansen B (1979) Is arterial hypertension crucial for the development of cerebral hemorrhage in premature infants? Lancet i: 1215–1217

244. Lou HC, Lassen NA, Tweed WA, Johnson G, Jones M, Palahmink RJ (1979) Pressure passive cerebral blood flow and break down of the blood-brain barrier in experimental fetal asphyxia. Acta Paediatr Scand 68: 57–63

245. Lou HC, Lassen NA, Friis-Hansen B (1979) Impaired autoregulation of cerebral blood flow in the distressed newborn infant. J Pediatr 94: 118–121

246. Lou HC, Skov H, Pedersen H (1979) Low cerebral blood flow: a risk factor in the neonate. J Pediatr 95: 606–609

247. Lou HC (1983) Perinatal cerebral ischaemia and developmental neurologic disorders. Act Paediatr Scand [Suppl] 311: 28–31

248. Lou HC, Tweed A (1983) Regulation of cerebral perfusion in utero. Neuropediatrics 14: 123 A

249. Lou HC, Tweed WA, Davies JM (1985) Preferential blood flow increase to the brain stem in moderate neonatal hypoxia: reversal by naloxone. Eur J Pediatr 144: 225–227

250. Lübbers DW (1972) Physiologie der Gehirndurchblutung. In: Gänshirt H (Hrsg) Der Hirnkreislauf. G Thieme, Stuttgart, S 214–254
251. Lütschg J, Hänggeli C, Huber P (1983) The evolution of cerebral hemispheric lesions due to pre-or perinatal asphyxia (clinical and neuroradiological correlation) Helv Paediat Acta 38: 245–254
252. Lundar T, Lindegaard KF, Froysaker T (1985) Cerebral perfusion during nonpulsatile cardiopulmonary bypass. Ann Thorac Surg 40: 144–150
253. Lundell BP (1983) Pulse wave patterns in patent ductus arteriosus. Arch Dis Childh 58: 682–685
254. Lundell BP, Lindstrom DP, Arnold TG (1984) Neonatal cerebral blood flow velocity I. Acta Paediat Scand 73: 810–815
255. Mannino FL, Trauner DA (1983) Stroke in neonates. J Pediatr 102: 605–610
256. Mantovani JF, Gerber GJ (1984) "Idiopathic" neonatal cerebral infarction. AJDC 138: 359–362
257. Markwalder TM, Grolimund P, Seiler RW, Roth F, Aaslid R (1984) Dependency of blood flow velocity in the middle cerebral artery on endtidal carbon dioxide partial pressure-a transcranial Doppler study. J Cerebral Blood Flow Metab 4: 368–372
258. Marshall TA, Marshall F, Reddy PP (1982) Physiologic changes associated with ligation of the ductus arteriosus in preterm infants. J Pediatr 101: 749–751
259. Marshall M (1984) Praktische Dopplersonographie. Springer, Berlin Heidelberg New York Tokyo
260. Martin CG, Snider AR, Katz SM, Peabody JL, Brady JP (1982) Abnormal cerebral blood flow patterns in preterm infants with large patent ductus arteriosus. J Pediatr 101: 587–593
261. Martin DJ, Hill A, Fitz CR, Daneman A, Havill DA, Becker LE (1983) Hypoxic/ischemic cerebral injury in the neonatal brain. Pediatr Radiol 13: 307–312
262. Matthes W, Dörstelmann D (1985) Klinischer und angiographischer Beitrag zum Moya-Moya-Syndrom. Act Neurol 12: 164–167
263. Mathew NT, Krastnik F, Meyer JS (1976) Regional cerebral blood flow in the diagnosis of vascular headache. Headache 15: 252–260
264. Mayersbach H (1956) Der Wandaufbau der Gefäßübergangsstrecken zwischen Arterien rein elastischen und rein muskulären Typs. Anat Anz 102: 333–360
265. McMenamin JB, Volpe JJ (1983) Doppler ultrasonography in the determination of neonatal brain death. Ann Neurol 14/3: 302–307
266. Mc Rae LP, Karcher M (1982) Oculoplethysmography. In: Kempczinski RF, Yao JST (eds) Practical noninvasive vascular diagnosis. Chicago Year Book, p 182
267. Menke JA, Miles R, McIlhany M, Bashiru M, Chua C, Schwied E, Menten TG, Khanna NN (1982) The fontanelle tonometer: A noninvasive method for measurement of intracranial pressure. J Pediatr 100: 960–963

268. Ment LR, Ehrenkrantz RA, Lange RC, Rothstein PT, Duncan CC (1981) Alterations in cerebral blood flow in preterm infants with intraventricular hemorrhage. Pediatrics 68: 763–769

269. Ment LR, Stewart WB, Lambrecht R, Duncan CC (1982) Local cerebral blood flow and metabolism in the beagle puppy model of intraventricular hemorrhage: an autoradiographic study. Ann Neurol 10: 289 A

270. Ment LR, Duncan CC, Ehrenkranz RA, Lange RC, Taylor KJ, Kleinman CS, Scott DT, Sivo J, Gettner P (1984) Intraventricular hemorrhage in the preterm neonate: Timing and cerebral blood flow changes. J Pediatr 104: 419

271. Meyer JE (1953) Über die Lokalisation frühkindlicher Hirnschäden in arteriellen Grenzgebieten. Arch Psych Nerv 190: 328–341

272. Miller GM, Black VD, Lubchenko LO (1981) Intracerebral hemorrhage in a term newborn with hyperviscosity. AJDC 135: 377–378

273. Milligan DWA (1980) Failure of autoregulation and intraventricular haemorrhage in preterm infants. Lancet i: 896–898

274. Miyazaki M, Kato K (1965) Measurement of cerebral blood flow by ultrasonic Doppler technique. Jpn Circ J 29: 375–382

275. Mochalowa LD, Khodov DA, Zhukova TP (1983) Cerebral circulation control in healthy full-term neonates. Acta Paediatr Scand [Suppl] 311: 20–22

276. Moscoso P, Goldberg RN, Jamielson J, Bancalari E (1983) Spontaneous elevation in arterial blood pressure during the first hours of life in the very-low-birth-weight infant. J Pediatr 103: 114–117

277. Moss A, Duffie ER, Emmanouilides G (1963) Blood pressure and vasomotor reflexes in the newborn infant. Pediatrics 32: 175–179

278. Mostafawy A (1971) Pediatric sonoencephalography. Springer, Berlin Heidelberg New York

279. Müller HR (1985) Quantitative Bestimmung des Blutflusses in der vena jugularis interna mittels Ultraschall. Ultraschall 6: 51–54

280. Mullaart RA, Krijgsman JB (1983) Ultrasound examination of the cerebral circulation in the newborn infant. Neuropediatrics 14: 123 A

281. Myerberg DZ, York C, Chaplin ER, Gregory GA (1980) Comparison of noninvasive and direct measurements of intracranial pressure. Pediatrics 65: 473–476

282. Myers RE (1977) Experimental models of perinatal brain damage: relevance to human pathology. In: Gluck L (ed) Intrauterine asphyxia and the developing fetal brain. Yearbook Medical Publishers, Chicago, pp 37–97

283. Namon R, Gollan F, Shimojyo S, et al (1967) Basic studies in rheoencephalography. Neurology 17: 239–252

284. Nathan DG, Oski FA (1981) Hematology of infancy and childhood. Saunders, Philadelphia London Toronto

285. Nelder JA, Mead R (1965) A simplex method for function minization. Comput J 7: 308–313

286. Neunzig HP, Einhäupl K, Hartmann A, Henze T (1985) Therapie der intrakraniellen Drucksteigerung. Act Neurol 12: 212–216

287. Newton TH, Potts DG (1971) Radiology of the skull and brain, vol 1. Mosby, St Louis

288. Nicolopoulos M, Perakis A, Papadakis M, Alexion D, Aravantinos D (1976) Estimation of gestational age in the neonate. AJDC 130: 477

289. Niederhoff H, Pringsheim W, Sutor AH, Hendrich G, Künzer W (1977) Hirnblutungen bei Neugeborenen. Monatsschr Kinderheilkd 125: 450–451

290. Nissen P (1986) Kritische Aspekte zur Dopplersonographie mit der Frequenzspektrumanalyse in der klinischen Anwendung. In: Otto RC, Schnaars P (Hrsg) Ultraschalldiagnostik 1985. G Thieme, Stuttgart

291. Nwasei C, Pape KE, Martin DJ, Becker LE, Fitz CR (1984) Periventricular infarction diagnosed by ultrasound: a postmortem correlation. J Pediatr 105: 106–110

292. Pape KE, Cusick G, Honang MTW, Blackwell RJ, Sherwood A, Thorburn RJ, Reynolds EOR (1979) Ultrasound detection of brain damage in preterm infants. Lancet i: 1261–1264

293. Pape KE, Wigglesworth JS (1979) Hemorrhage, ischaemia and the perinatal brain. Heinemann, London

294. Papile LA, Burstein J, Burstein R, Koffler H (1978) Incidence and evolution of subependymal and intraventricular hemorrhage: a study of infants with birth weight less than 1,500 g. J Pediatr 92: 529–534

295. Papile LA, Rudolph AM, Heymann MA (1982) Autoregulation of cerebral blood flow in the preterm ovine fetus. Pediatr Res 16: 339 A

296. Papile LA, Rudolph AM, Heymann MA (1985) Autoregulation of cerebral blood flow in the preterm fetal lamb. Pediatr Res 19: 159–161

297. Pasternak JF, Groothuis DR, Fisher JM, Fisher DP (1982) Regional cerebral blood flow in the newborn beagle pup: the germinal matrix is a "low-flow" structure. Pediatr Res 16: 499–503

298. Pasternak JF, Groothuis DR (1985) Autoregulation of cerebral blood flow in the newborn beagle puppy. Biol Neonate 48: 100–109

299. Patel A, Toole JF (1965) Subclavian steal syndrome-reversal of cephalic blood flow. Medicine 44: 289

300. Peabody JL (1981) Muscle relaxants-a potential danger to infants at risk for intraventricular hemorrhage. Pediatr Res 15: 709 A

301. Perlman JM, Hill A, Volpe JJ (1981) The effect of patent ductus arteriosus on the flow velocity in the anterior cerebral arteries: Ductal steal in the premature newborn infant. J Pediatr 99: 767

302. Perlman JM, Volpe JJ (1982) Cerebral blood flow velocity in relation to intraventricular hemorrhage in the premature newborn infant. J Pediatr 100: 956–959

303. Perlman JM, Volpe JJ (1983) Seizures in the preterm infant: Effect on cerebral blood flow velocity, intracranial pressure, and arterial blood pressure. J Pediatrics 102: 288–293

304. Perlman JM, Volpe JJ (1983) Suctioning in the preterm infant: Effects on cerebral blood flow velocity, intracranial pressure and arterial blood pressure. Pediatrics 72: 329–334

305. Perlman JM, McMenamin JB, Volpe JJ (1983) Fluctuating cerebral blood-flow velocity in respiratory distress syndrome. N Engl J Med 309: 204–209

306. Perlman JM, Volpe JJ (1985) Episodes of apnea and bradycardia in the preterm newborn: impact on cerebral circulation. Pediatrics 76: 333–338

307. Pfenninger J (1984) Neurointensivpflege im Kindesalter. Huber, Bern Stuttgart Toronto

308. Philipp AGS, Long JG, Donn SM (1981) Intracranial pressure. Sequential measurements in full-term and preterm infants. Am J Dis Child 135: 521–524

309. Planiol T, Pourcelot L (1974) Doppler effect study of the carotid circulation. In: De Vlieger M, White DN, Mc Cready VR (eds) Second world congress on ultrasonics in medicine. Excerpta Medica, Amsterdam, pp 104–111

310. Portnoy HD, Brauch C, Chopp M (1985) The CSF pulse wave in hydrocephalus. Child's Nerv Syst 1: 248–254

311. Pourcelot L (1975) Applications cliniques de l'examen Doppler transcutane. In: Peronneau P (ed) Velocimétrie ultrasonore Doppler. INSERM, Paris, pp 213–224

312. Pourcelot L, Arbeille P, Pottier JM, Patat F, Mignier P, Guell A, Gharib C (1984) Ultrasonic study of early cardiovascular adaptation to zero gravity. Proceedings, 2nd european symposium on life science; research in space, Porz

313. Prechtl HFR, Beintema DJ (1976) Die neurologische Untersuchung des reifen Neugeborenen, 2. Aufl. G Thieme, Stuttgart

314. Rahilly PM (1980) Effects of sleep state and feeding on cranial blood flow of the human neonate. Arch Dis Childh 55: 265–270

315. Raju TNK, Vidyasagar D (1980) Intracranial pressure studies in acutely ill neonates using a noninvasive technique: a 3 year experience. In: Shulman J (ed) Intracranial pressure, vol 4. Springer, Berlin Heidelberg New York, pp 392–394

316. Raju TNK, Doshi UV, Vidyasagar D (1982) Cerebral perfusion pressure studies in healthy preterm and term newborn infants. J Pediatr 100: 139–142

317. Raju TNK, Doshi UV, Vidyasagar D (1983) Low cerebral perfusion pressure: an indicator of poor prognosis in asphyxiated term infants. Brain Develop (Int Ed) 5: 478–482

318. Rautenberg W, Hennerici M (1986) Transkranielle Dopplersonographie bei asymptomatischen Patienten mit hochgradigen extrakraniellen Strömungsbehinderungen. In: Otto RC, Schnaars P (Hrsg) Ultraschalldiagnostik 1985. G Thieme, Stuttgart New York

319. Reivich M (1964) Arterial pCO_2 and cerebral hemodynamics. Am J Physiol 206: 25

320. Revich M (1972) Zerebrale Autoradiographie. In: Gänshirt H (Hrsg) Der Hirnkreislauf. G Thieme, Stuttgart

321. von Reutern G, Pourcelot L (1978) Cardiac cycle dependent alternating flow in vertebral arteries with subclavian artery stenoses. Stroke 9: 229

322. von Reutern GM, Büdingen HJ (1981) Möglichkeiten und Grenzen der Dopplersonographie an den extrakraniellen Hirnarterien. Ultraschall 2: 35–42

323. von Reutern GM, Arnolds B (1985) Transcranial Doppler sonography in cerebrovascular disease. Stroke 16: 16A

324. Rickenbacher J (1972) Normale und pathologische Anatomie des Hirngefäß-systems. In: Gänshirt H (Hrsg) Der Hirnkreislauf. G Thieme, Stuttgart, S 25–30

325. Ringelstein EB, Wulfinghoff F, Zeumer H, Grosse W, Korbmacher G (1985) Diagnostische Möglichkeiten der transkraniellen Dopplersonographie in der Neurologie. Angio 7: 167–182

326. Ringelstein EB (1985) Ultraschalldiagnostik am vertebrobasilären Kreislauf. Teil II: Transnuchale Diagnose intrakranieller vertebrobasilärer Stenosen mit Hilfe eines neuartigen Impulsschall-Doppler-Systems. Ultraschall 6: 60–67

327. Ringelstein EB, Zeumer H, Korbmacher G, Wulfinghoff F (1985) Transkra-nielle Dopplersonographie der hirnversorgenden Arterien: atraumatische Diagnostik von Stenosen und Verschlüssen des Carotissiphons und der A. cerebri media. Nervenarzt 56: 296–306

328. Ritchie WL, Overton TR (1980) Vasospasm and cerebral blood flow after subarachnoid hemorrhage. J Can Ass Radiol 31: 230–233

329. Ritchie WL, Overton TR (1981) Vasospasm and cerebral blood flow after subarachnoid hemorrhage. AJR 136: 1263

330. Rolak LA, Rokey R (1986) Magnetic resonance imaging in Moya-Moya disease. J Child Neurol 1: 67–70

331. Rolfe P, Persson B, Zetterström R (1983) An appraisal of techniques for studying cerebral circulation in the newborn. Act Paediatr Scand [Suppl] 311: 5–13

332. Rosenberg AA, Narayanan V, Jones D (1985) Comparison of anterior ce-rebral artery blood flow velocity and cerebral blood flow during hypoxia. Pediatr Res 19: 67–70

333. Rosenblum WI, Asofsky RM (1968) Malfunction of cerebral microcirculation in macroglobulinemic mice-relationship to increased blood viscosity. Arch Neurol 18: 151–159

334. Rosenfeld W, Sandhev S, Brunot V, Ihaveri R, Zabelata I, Evans HE (1986) Phototherapy effect on the incidence of patent ductus arteriosus in premature infants: Prevention with chest shielding. Pediatrics 78: 10–14

335. Rosenkrantz TS, Oh W (1982) Cerebral blood flow in infants with polycy-themia and hyperviscosity: Effects of partial exchange transfusion with plas-manate. J Pediatr 101: 94–98

336. Rosenkrantz TS, Oh W (1984) Aminophylline reduces cerebral blood flow velocity in low-birth-weight infants. AJDC 138: 489–491

337. Rott HD (1988) Bioeffect. Springer, Berlin Heidelberg New York Tokyo

338. Sankaran K, Peters K, Finer N (1981) Estimated cerebral blood flow in term infants with hypoxic-ischemic encephalopathy. Pediatr Res 15: 1415–1418

339. Saternus KS (1985) Bauchlage im Neugeborenen- und Säuglingsalter. Päd Praxis 32: 305–307

340. Satomura S (1959) Study on the flow patterns in peripheral arteries by ultrasonics. J Acoust Soc Jpn 15: 151–158

341. Salmon JH, Hajjar W, Bada HS (1977) The fontanogram: A noninvasive intracranial pressure monitor. Pediatrics 60: 721–725

342. Savage JM, Dillon MJ, Taylor JFN (1979) Clinical evaluation and com-

parison of the infrasond arteriosonde, and mercury sphygmomanometer in measurement of blood pressure in children. Arch Dis Childh 54: 184–189

343. Scheffner D, Wille L (1973) Acute infantile hemiplegia due to obstruction of intracranial vessels. Neuropädiatrie 4: 7–19

344. Schiefer W, Vetter K (1957) Das zerebrale Angiogramm in den verschiedenen Altersstufen. Zbl Neurochir 17: 218–231

345. Schimmer M (1982) Hirnödem beim diabetischen Koma im Kindesalter. DMW 107: 1111–1113

346. Schmidt K (1972) Hirndurchblutung bei intrakranieller Drucksteigerung und beim Hirnödem. In: Gänshirt H (Hrsg) Der Hirnkreislauf. G Thieme, Stuttgart, S 715–729

347. Schöber JG (1986) Perinatale Adaptation von Herz und Kreislauf und ihre Störungen. In: Neuhäuser G (Hrsg) Entwicklungsstörungen des Zentralnervensystems. Kohlhammer, Stuttgart

348. Seiler R, Aaslid R (1986) Transcranial Doppler for evaluation of cerebral vasospasm. In: Aaslid R (ed) Transcranial Doppler sonography. Springer, Wien New York

349. Serwer GA, Armstrong BE, Anderson PAW (1980) Noninvasive detection of retrograde descending aortic flow in infants using continous wave Doppler ultrasonography. J Pediatr 97: 394

350. Serwer GA, Armstrong BE, Sterba RJ, et al. (1981) Alterations in carotid arterial velocity-time profile produced by the Blolock-Taussig shunt. Circulation 63: 1115

351. Serwer GA, Armstrong BE, Anderson PA (1982) Continous wave Doppler ultrasonographic quantitation of patent ductus arteriosus flow. J Pediatr 100: 297

352. Seydel HG (1964) The diameters of the cerebral arteries of the human fetus. Anat Rec 150: 79–86

353. Shenkin HA, Spitz EB, Grant FC, Kety SS (1948) The acute effects on the cerebral circulation of the reduction of increased pressure by means of intravenous glucose or ventricular drainage. J Neurosurg 5: 466–501

354. Shulman J (Hrsg) (1980) Intracranial pressure, vol 4. Springer, Berlin Heidelberg New York, pp 392–394

355. Siegel MJ, Shackelford GD, Perlman JM, Fulling KH (1984) Hypoxic-ischemic encephalopathy in term infants: Diagnosis and prognosis evaluated by ultrasound. Radiology 152: 395–399

356. Skinhoj E (1973) Hemodynamic studies within the brain during migraine. Arch Neurol 29: 95–98

357. Slovis TL, Shankaran S (1984) Ultrasound in the evaluation of hypoxic-ischemic injury and intracranial hemorrhage in neonates: the state of the art. Pediatr Radiol 14: 67–75

358. Smith BT, Boyle JM, Dongarra JJ, Garbow BS, Ikebe Y, Klema VC, Moler CB (1976) Matrix eigensystem routines-EISPACK-guide, 2nd edn. Springer, Berlin Heidelberg New York

359. Sokoloff LB (1978) Postnatal maturation of the local cerebral circulation. In: Berenberg SR (ed) Brain, fetal and infant-current research on normal and abnormal development. Nijhoff, Medical Division, The Hague

360. Spencer MP (1983) Intracranial carotid artery diagnosis with transorbital pulsed wave (PW) and continous wave (CW) Doppler ultrasound. J Ultrasound Med [Suppl] 2: 61

361. Spendley W, Hext GR, Himsworth FR (1962) Sequential application of simple designs in optimisation and evolutionary operating. Technometrics 4: 441

362. Stave U (ed) (1978) Perinatal physiology, 2nd edn. Plenum Medical Book Company, New York London

363. von Stockhausen HB, Bopp E, Mahdi S (1984) Fontanellometrie als Monitoring einer möglichst schonenden Pflege und Therapie sehr kleiner Frühgeborener. In: Kowalewki S (Hrsg) Pädiatrische Intensivmedizin, Bd 6. G Thieme, Stuttgart New York, S 47–48

364. Stopfkuchen H, Schranz D, Schwarz M, Tegtmeyer F, Wirth S (1984) Behandlungsschema bei Kindern mit schwerem Schädelhirntrauma nach Einführung der intrakraniellen Druckmessung. Monatsschr Kinderheilkd 132: 58–61

365. Straßburg HM, Niederhoff H, Sauer M (1982) Die Dopplersonographische Registrierung der Durchblutung intakranieller Gefäße beim Säugling. Monatsschr Kinderheilkd 130: 608–612

366. Straßburg HM, Sauer M (1982) Morphologische Darstellung und Identifizierung eines Aneurysma der Vena Galeni beim Säugling mit der Duplex-Scan-Technik. Klin Pädiatr 194: 84–87

367. Straßburg HM, Pringsheim W, Bohlayer R (1982) Diagnostik generalisierter und fokaler Hirnödeme mittels der Duplex-Scan-Technik beim Neugeborenen. In: Saling E, Dudenhausen J (Hrsg) Perinatale Medizin, Bd 9. G Thieme, Stuttgart New York, S 343–344

368. Straßburg HM, Bohlayer R, Abel M (1984) Die Dopplerfrequenzanalyse intrakranieller Gefäße beim Neugeborenen mit Hirnblutung. In: Kowalewski S (Hrsg) Pädiatrische Intensivmedizin, Bd 6. G Thieme, Stuttgart New York

369. Straßburg HM, Bogner K (1986) Änderungen der zerebralen Durchblutung beim gesunden Neugeborenen in den ersten drei Lebenstagen. In: Neuhäuser G (Hrsg) Entwicklungsstörungen des Zentralnervensystems. Kohlhammer, Stuttgart

370. Straßburg HM, Klemm JF, Wais U, Göppinger A (1984) Nichtinvasive Hirndruckmessung über der vorderen Fontanelle bei gesunden Säuglingen in den ersten Lebenstagen. Monatsschr Kinderheilkd 132: 904–908

371. Straßburg HM, Bode H, Dahmen U (1986) Der prognostische Wert der zerebralen Sonographie. Klin Pädiatr 198: 385–390

372. Stuart B, Drumm J, Fitzgerald DE, Duignan NM (1980) Fetal blood velocity waveforms in normal pregnancy. Br J Obstet Gynaecol 87: 780

373. Stullken EH, Johnston WE, Prough DS, Babstrieri FJ, McWhorster JM (1985) Implications of nimodipine prophylaxis of cerebral vasospasm on anaesthetic management during intracranial aneurysm clipping. J Neurosurg 62: 200–205

374. Surry L (ed) European Medical Ultrasonics 7/1: 4–10

375. Surry L (ed) European Medical Ultrasonics 7/2: 3–11

376. de Swiet M, Fayers P, Shinebourne EA (1980) Systolic blood pressure in a

population of infants in the first year of life: The Brompton study. Pediatrics 65: 1028–1035

377. Suzuki J, Takaku A (1969) Cerebrovascular moyamoya disease. Arch Neurol 20: 288–299

378. Szymonowicz W, Yu VYH, Wilson FE (1984) Antecedents of periventricular haemorrhage in infants weighing 1,250 g or less at birth. Arch Dis Childh 59: 13–17

379. Szymonowicz W, Yu VYH (1984) Timing and evolution of periventricular haemorrhage in infants weighing 1,250 g or less at birth. Arch Dis Childh 59: 7–12

380. Szymonowicz W, Yu VYH, Walker A, Wislon F (1986) Reduction inter-ventricular haemorrhage in preterm infants. Arch Dis Childh 61: 661–665

381. Thomas DJ, Marshall J, Ross Russell RW, Wetherley-Mein G, duBoulay GH, Pearson TC, Symon L, Zilkja E (1977) Effect of haematocrit on blood flow in man. Lancet ii: 941

382. Trowitzsch E, Burger BM, Sanders SP (1986) Erfahrungen mit der zweidi-mensionalen Dopplerechokardiographie bei Kindern mit angeborenen und erworbenen Herzfehlern. Monatsschr Kinderheilkd 134: 67–75

383. Tweed WA, Cote J, Wade JG, Gregory G, Mills A (1982) Preservation of fetal brain blood flow relative to other organs during hypovolemic hypo-tension. Pediatr Res 16: 137–140

384. Uematsu S, Yang A, Preziosi TJ, Kouba R, Tong TJK (1983) Measurement of carotid blood flow in man and its clinical application. Stroke 14: 256–266

385. Vanucci RC, Plum F (1976) Pathophysiology of perinatal hypoxic ischemic brain damage. In: Gaull GE (ed) Biology of brain dysfunction, vol 3. Plenum, New York

386. Vanucci RC, Hernandez MJ (1979) Cerebral blood flow re intracranial hem-orrhage. J Pediatr 95: 496–498

387. Vergesslich KA, Weninger M, Simbruner G, Ponhold W (1986) Beziehungen zwischen gepulster und kontinuierlicher Doppler-Velocimetrie der Arteria cerebri anterior beim Neugeborenen. 35. Jahrestagung der Süddeutschen Gesellschaft für Kinderheilkunde, Augsburg, 6–8. 6. 1986

388. Versmold HT, Kitterman JA, Phibbs RH, Gregory GA, Tooley WH (1981) Aortic blood pressure during the first 12 hours of life in infants with birthweight 610–4,220 grams. Pediatrics 67: 607–613

389. Vetter K (1986) Die intrauterine Blutflußmessung beim Feten mittels Dopp-ler-Ultraschall. In: Otto RC, Schnaars P (Hrsg) Ultraschalldiagnostik 1985. G Thieme, Stuttgart New York

390. Vidyasagar D, Raju TNK (1977) A simple noninvasive technique of mea-suring intracranial pressure in the newborn. Pediatrics 59: 957–962

391. Vidyasagar D, Raju TNK, Chiang J (1978) Clinical significance of monitoring anterior fontanelle pressure in sick neonates and infants. Pediatrics 62: 996–999

392. Vogel M, Dörlemann C, Weil J (1986) Prospektive zweidimensionale Echocardiographie zur Untersuchung des Ductus arteriosus (PDA) bei beatmeten Frühgeborenen. Monatsschr Kinderheilkd 134: 602A
393. Voigt K, Brand T, Sauer M (1972) Röntgenanatomische Variationsstatistik zur topographischen Beziehung zwischen A. basilaris und Schädelbasisstrukturen: Neuroradiologische Untersuchungen an Vertebralis- und Brachialisangiographien. Arch Psychiat Nervenkr 215: 376–395
394. Voigt K, Stoeter P (1980) Neuroradiologie der embryonalen Hirnentwicklung. Enke, Stuttgart
395. Volpe JJ (1978) Neonatal periventricular hemorrhage: past, present, future. J Pediatr 92: 693–696
396. Volpe JJ (1979) Cerebral blood flow in the newborn infant: relation to hypoxic-ischemic brain injury and periventricular hemorrhage. J Pediatr 94: 170–173
397. Volpe JJ (1981) Current concepts in neonatal medicine. Neonatal intraventricular hemorrhage. N Engl J Med 304: 886–891
398. Volpe JJ (1987) Neurology of the newborn. 2nd edn. Saunders, Philadelphia
399. Volpe JJ, Perlman JM, Hill A, Mc Menamin JB (1982) Cerebral blood flow velocity in the human newborn: the value of its determination. Pediatrics 70: 147–152
400. Volpe JJ (1982) Anterior fontanel: window to the neonatal brain. J Pediatr 100: 395–398
401. Volpe JJ, Herscovitch P, Perlman JM, Raichle ME (1983) Positron emission tomography in the newborn: extensive impairment of regional cerebral blood flow with intraventricular hemorrhage and hemorrhagic intracerebral involvement. Pediatrics 72: 589–601
402. Voorhies TM, Lipper EG, Lee BCP, Vanucci RC, Auld PAM (1984) Occlusive vascular disease in asphyxiated newborn infants. J Pediatr 105: 92–96
403. deVries LS, Dubowitz LMS, Dubowitz V et al. (1985) Predictive value of cranial ultrasound in the newborn baby: a reappraisal. Lancet ii: 137–140
404. Walsh SZ, Lind J (1978) The fetal circulation and its alteration at birth. In: Stave U (ed) Perinatal physiology. Plenum Medical Book Company, New York London, pp 129–180
405. Walsh P, Logan WJ (1983) Continous and intermittent measurement of intracranial pressure by LADD monitor. J Pediatr 102: 439–442
406. Wassmann H (1986) Nimodipin löst Hirngefäßspasmen. Arzneimitteltherapie 4: 111
407. Weindling AM, Rolfe P, Tarassenko L, Costeloe K (1983) Cerebral hemodynamics in newborn babies studied by electrical impedance. Acta Paediatr Scand [Suppl] 311: 14–19
408. Welch K (1980) The intracranial pressure in infants. J Neurosurg 52: 693–699
409. Wenner J (1972) Entwicklung der Kapillarisierung und der Sauerstoffversorgung des Gehirns im Säuglingsalter. In: Gänshirt H (Hrsg) Der Hirnkreislauf. G Thieme, Stuttgart, S 201–212
410. Wenner J (1964) Über die Entwicklung des O_2-Verbrauchs und der Durch-

blutung des Gehirns im Säuglingsalter. Monatsschr Kinderheilkd 112: 242–244
411. Westphal W (1970) Physik, 26. Aufl. Springer, Berlin Heidelberg New York
412. Widder B (1985) Der CO_2-Test zur Erkennung hämodynamisch kritischer Carotisstenosen mit der transkraniellen Doppler-Sonographie. DMW 110: 1553
413. Wiese G (1983) Hyperbilirubinämie des Neugeborenen. Monatsschr Kinderheilkd 131: 193–203
414. Wilcox WD, Carrigan TA, Dooley KJ, Giddens DP, Dykes FD, Lazzara A, Ray JL, Ahmann PA (1983) Range-gated pulsed Doppler ultrasonographic evaluation of carotid arterial blood flow in small preterm infants with patent ductus arteriosus. J Pediatr 102: 294–298
415. Wilcox WD, Carrigan TA, Ahmann PA (1983) Direction of flow with PDA-Reply. J Pediatr 103: 833
416. Wimberley PD, Lou HC, Pedersen H, Hejl M, Lassen NA, Friis-Hansen B (1982) Hypertensive peaks in the pathogenesis of intraventricular hemorrhage in the newborn. Abolition by phenobarbitone sedation. Acta Paediatr Scand 71: 537–542
417. Winter R, Hohagen F, Kaiser W, Reuther R (1987) Reproduzierbarkeit transkranieller dopplersonographischer Messungen. In: Widder B (Hrsg) Transkranielle Doppler-Sonographie bei zerebrovaskulären Erkrankungen. Springer, Berlin Heidelberg New York Tokyo
418. Wiswell TE, Cornish JD, Northam RS (1986) Neonatal polycythemia: frequency of clinical manifestations and other associated findings. Pediatrics 78: 26–30
419. Wozniak M, Mc Lone DG, Raimondi AJ (1975) Micro- and macrovascular changes as the direct cause of parenchymal destruction in congenital murine hydrocephalus. J Neurosurg 43: 535–545
420. Yamashima T, Kashihara K, Ikeda K, Kubota T, Yamamoto S (1985) Three phases of cerebral arteriopathy in meningitis: vasospasm and vasodilation followed by organic stenosis. Neurosurgery 16/4: 546–553
421. Ylppö A (1924) Zum Entstehungsmechanismus von Hirnblutungen beim Frühgeborenen. Z Kinderheilkd 38: 32–45
422. Young RSK, Hernandez MJ, Yagel SK (1982) Selective reduction of blood flow to white matter during hypotension in newborn dogs: possible mechanism of periventricular leucomalacia. Ann Neurol 12: 445–449
423. Younkin DP, Reivich M, Obrist W, Delivoria-Papadopulos M (1981) Non-invasive neonatal regional cerebral blood flow. Pediatr Res 15: 713A

Subject Index

144

Subject Index

Position, of head and body 29, 84
Positron-emission tomography 10
Pressure passivity 51, 101, 103
Pressure waves 56, 100
Prognosis 111
Prophylaxis 100
Pulmonary atresia 49
Pulsatility index 14, 76, 77, 81

Reference values 44, 93
REM sleep 7, 86
Reproducibility 27, 82, 83
Resistance index 14, 76–77, 81, 103
Results 25–75
Reverberating flow pattern 51, 64, 98, 103
Reynold's number 5

Sample volume 13, 16, 82
SD-ratio 14, 30, 34, 47, 49, 55, 59, 74, 77, 88
Seizure 59, 99, 103
Sepsis 74, 107
Sickle cell anemia 110
Sides, comparison of 28, 83
Skull diameter 25
SM-ratio 14, 48, 77, 94
Spherocytosis 110
Stenosis, vascular 67, 79, 92, 104–107, 108
Subarachnoid hemorrhage 97, 105
Subclavian steal syndrome 95

Subdural hematoma 59, 69, 103, 105
Suctioning 99

Therapy control 75, 99, 108
Thromboemboly 96
Transcranial Doppler sonography 12, 13, 16, 18, 25, 82, 108–111
continuous recording 110
procedure 22–24
recording techniques 8–22, 82, 109
three-dimensional 21, 83

Ultrasound
acoustic intensity 16
acoustic windows 13, 18
A-scan 11
B-scan 11, 24, 64, 104
emitting frequency 12, 81
focus 13
insonation time 10
power 10

Valsalva test 56, 100
Variability
interindividual 30
intraindividual 27, 83
Vasospasm 68–69, 105–107
Veins 6, 100, 101, 111
Vessel diameter 5, 78–79, 84, 87
Vigilance 30, 85–86
Viscosity 6, 79, 88
Volume flow 6, 77, 84

A. Harders

Neurosurgical Applications of Transcranial Doppler Sonography

1986. 109 figures. X, 134 pages. ISBN 3-211-81938-X
Soft cover DM 58,–, öS 406,–
Prices are subject to change without notice

In 1981 Dr. Rune Aaslid developed a transcranial Doppler device with a pulsed sound emission of 2 MHz, which enabled blood flow velocities to be measured in the large branches of the circle of Willis. With this innovation, it has become possible to record atraumatically and repeatedly the intracranial hemodynamic changes in neuro-vascular diseases.

The book describes the hemodynamic principles in cerebral vascular circulation and the factors which can effect the blood flow velocities (such as collateral circulation, diameter of the vessels, vascular resistance, arterial partial CO_2 pressure, autoregulatory factors, and position of the body). Normal values of blood flow velocities and the changes under physiological deviations are measured by transcranial Doppler technique. For patients suffering from subarachnoid hemorrhage, individual time courses of velocity changes are evaluated and the application in clinical routines is stressed: Better defined timing of angiography, surgery and postoperative hypertension therapy has significantly reduced the incidence of delayed ischemic deficits. Patients indicating for extracranial-intracranial bypass surgery, as well as the postoperative changed hemodynamics are also investigated. The contribution of the bypass to the brain circulation can be tested by compression tests. The "activity" of an angioma and the influence of superselective embolization procedures for arteriovenous malformations are described.

Furthermore, cerebro-vascular blood flow arrest in brain death patients, can clearly be seen without angiography by evaluating a reverberating flow pattern. The book gives an account of the role of a still very young but exciting technique in diagnostic and therapeutic procedures of cerebral vascular disease based upon three years of experience at the Neurosurgical Department of the University of Freiburg.

Springer-Verlag Wien New York

Springer-Verlag, Moelkerbastei 5, A-1010 Wien;
175 Fifth Avenue, New York, NY 10010, USA;
Heidelberger Platz 3, D-1000 Berlin 33;
37-3, Hongo 3-chome, Bunkyo-ku, Tokyo, Japan

Transcranial Doppler Sonography

Edited by R. Aaslid

1986. 94 figures. XI, 177 pages. ISBN 3-211-81935-5
Soft cover DM 68,–, öS 476,–

Prices are subject to change without notice

From the Foreword by M.P. Spencer, M.D., Director of the Institute of Applied Physiology and Medicine, Seattle, Washington, U.S.A.:
"Every few years a dissertation comes to the area of clinical application of medical technology which carries us forward as on a magic carpet into new regions of understanding and patient care. This book is such a magic carpet. It brings together, in a clear and incisive fashion, important hemodynamic principles with a simple non-invasive method of application to a part of the cerebral vasculature which has been relatively inaccessible. To the lucky and perceptive person who reads this book, a feeling of excitement and hope for progress is engendered. The diligent application of the potentials of transcranial Doppler ultrasound brings new power to our efforts in understanding the cerebral circulation and the causes, treatment and prevention of cerebrovascular disorders."

Springer-Verlag Wien New York

Moelkerbastei 5, A-1010 Wien;
175 Fifth Avenue, New York, NY 10010, USA;
Heidelberger Platz 3, D-1000 Berlin 33;
37-3, Hongo 3-chome, Bunkyo-ku, Tokyo, Japan